T0136340

Analytics and Control

Series Editor:
Adedeji B. Badiru
Air Force Institute of Technology, Dayton, Ohio, USA

Mechanics of Project Management
Nuts and Bolts of Project Execution
Adedeji B. Badiru, S. Abidemi Badiru, and I. Adetokunboh Badiru

The Story of Industrial Engineering
The Rise from Shop-Floor Management to Modern Digital Engineering
Adedeji B. Badiru

Innovation
A Systems Approach
Adedeji B. Badiru

For more information about this series, please visit: https://www.crcpress.com/
Analytics-and-Control/book-series/CRCAC

Innovation
A Systems Approach

Adedeji B. Badiru

CRC Press
Taylor & Francis Group
Boca Raton London New York

CRC Press is an imprint of the
Taylor & Francis Group, an **informa** business

First edition published 2020
by CRC Press
6000 Broken Sound Parkway NW, Suite 300, Boca Raton, FL 33487-2742

and by CRC Press
2 Park Square, Milton Park, Abingdon, Oxon, OX14 4RN

© 2020 Taylor & Francis Group, LLC

CRC Press is an imprint of Taylor & Francis Group, LLC

Library of Congress Cataloging-in-Publication Data

Names: Badiru, Adedeji Bodunde, 1952- author.
Title: Innovation : a systems approach / by Adedeji Badiru.
Description: First edition. | Boca Raton, FL : CRC Press, 2020. |
Series: Analytics and control | Includes bibliographical references.
Identifiers: LCCN 2019056660 | ISBN 9780367190859 (hardback) |
ISBN 9780429200281 (ebook)
Subjects: LCSH: Systems engineering. | Project management. |
Problem solving. | Technological innovations.
Classification: LCC TA168 .B24 2020 | DDC 658.4/032—dc23
LC record available at https://lccn.loc.gov/2019056660

ISBN: 978-0-367-19085-9 (hbk)
ISBN: 978-0-429-20028-1 (ebk)

Typeset in Times
by codeMantra

Dedicated to the memory of late Dr. Gary E. Whitehouse, my dearly-departed doctoral advisor, who originally introduced me to the concept of Systems Innovation in 1982 at the University of Central Florida.

Contents

Preface

It is a systems world. This book uses a systems-based approach to show how innovation is pervasive in all facets of endeavors, including business, industry, government, the military, and even academia. The systems approach facilitates process design, evaluation, justification, and integration. A key power of the book is that it explicitly highlights the eventual role of integration in all innovation implementation goals. Innovation is the lifeline of national development. This book presents chapters that provide techniques and methodologies for achieving the transfer of science and technology assets for innovation applications. Innovation can mean different things to different people. This book presents conceptual and operational definitions of innovation. Emphasis is placed on the context related to the theme of systems thinking. Some definitions are used within the framework of conceptual processes, while some are used within the platform of technology. An organization can be innovative in process-related activities without being involved in technological innovation. For this reason, a clarification of the various meanings and contexts of innovation is essential. This book meets that purpose. Foremost in this process of innovation is the role of organizational leadership in the pursuit of innovation in all its ramifications.

Adedeji B. Badiru
20 January 2020

Acknowledgments

Special thanks to Cindy Carelli and her team at the CRC Press/Taylor & Francis Group for the nurturing nudging toward more knowledge sharing through the introduction of new books. My gratitude also goes to my students and colleagues, whose new insights guide where each new book should head.

Author

Prof. Adedeji B. Badiru is dean and senior academic officer for the Graduate School of Engineering and Management at the Air Force Institute of Technology (AFIT). He was previously professor and head of Systems Engineering and Management at the AFIT, professor and department head of Industrial & Information Engineering at the University of Tennessee in Knoxville, and professor of Industrial Engineering and dean of University College at the University of Oklahoma, Norman. He is a registered professional engineer (PE), a certified project management professional (PMP), a Fellow of the Institute of Industrial Engineers, and a fellow of the Nigerian Academy of Engineering. He holds BS in Industrial Engineering, MS in Mathematics, and MS in Industrial Engineering from Tennessee Technological University, and PhD in Industrial Engineering from the University of Central Florida. His areas of interest include mathematical modeling, project modeling and analysis, economic analysis, systems engineering, and efficiency/productivity analysis and improvement. He is the author of over 25 books, 34 book chapters, 70 technical journal articles, and 110 conference proceedings and presentations. He has also published 25 magazine articles and 20 editorials and periodicals. He is a member of several professional associations and scholastic honor societies. He is the series editor for Taylor & Francis Group Systems Innovation series as well as the Analytics and Control series.

Systems Principles of Innovation

1

BACKDROP

On the platform of innovation, everything is interconnected in a system of systems. Innovation has always been embedded in human endeavors. Even though we are experiencing a global push for innovation now, it has always been prevalent in human's operational endeavors. Even in prehistoric times, there were increments of what we could call innovation in that era, particularly in progressive development of household (or "hut-hold") tools and hunting implements. In the context of our modern society, innovation, beyond its mere literary meaning, entails a multifaceted coordination of people, resources, budget, and organizational intellectual assets. For this reason, this book uses an integrative methodology of systems thinking to advance the narrative of innovation. A common fallacy of innovation is that innovation is technology based. Although technology innovation is often the most readily seen and appreciated, innovation can be of any of the following:

- Technology innovation
- Process innovation
- People (workforce) innovation
- Financial innovation.

Essentially, innovation should be viewed broadly as the act of leveraging whatever assets are available to an organization or an individual. There is a wave of interest in innovation raging through business, industry, academia, government, and the military. The emergence of the new digital economy has necessitated viewing innovation as not just a word but as an organizational asset requiring explicit managerial coordination. Every major undertaking by large organizations seems to be prefaced with the word "innovation." The fast pace of new technology is predicated on innovation. The era of smartphones is a great example of the rapidity of introducing new products. Many times, inventors, investors, and entrepreneurs innovate without understanding the full potential of what innovation entails and how to manage it.

The fact that "innovation" is not a tangible entity means that it is often taken for granted and poorly executed. Using a systems approach, innovation can be more easily embraced, actualized, and sustained.

INNOVATION REVOLUTION

Suddenly, everyone is talking innovation as if it is a new phenomenon. But innovation has always been with the human kind. It has existed since prehistorical times when humans learned how to survive by hunting and gathering. Although innovations of today are targeted more toward the needs of business and industry, the precursors to the present wave are the human's drive and strive to do things better, more efficiently, and more effectively over the years. What we are experiencing in the modern era is the heightened consciousness to coalesce efforts to do new things using new tools and techniques of today in preparation for tomorrow.

DEFINITIONS OF INNOVATION

In the final analysis, innovation is nothing more than the creation of an enabling environment that facilitates taking advantage of the capabilities of people, process, and technology.

It is a systems world. This book uses a systems-based approach to show how innovation is pervasive in all facets of endeavors, including business, industry, government, the military, and even academia. The systems approach facilitates process design, evaluation, justification, and integration (DEJI). A key power of the book is that it explicitly highlights the ultimate role of integration in all innovation implementation goals. Innovation is the lifeline of national development. This book presents chapters that provide techniques and methodologies for achieving the transfer of science and technology assets for innovation applications. Innovation can mean different things to different people. This chapter presents conceptual and operational definitions of innovation. Emphasis is placed on the context related to the theme of systems thinking. Some definitions are used within the framework of conceptual processes, while some others are used within the platform of technology. An organization can be innovative in process-related activities without being involved in technological innovation. For this reason, a clarification of the various meanings and contexts of innovation is essential.

The term "innovation" has become the word of the day in business, industry, government, and the military. Everyone talks about innovation, but very few appreciate the mission-dependent implications. Even the proponents of innovation often wonder what innovation implies. For this reason, this chapter considers alternate definitions, views, and concepts underpinning innovation. Innovation is not necessarily a tangible product that can be physically appraised. Unfortunately, this is the default assumption of those who tout innovation. Such a default view of innovation as an end product misses the opportunity to recognize the necessary interplay of technology, people, and process. Innovation is a process rather than a product. Any methodology (of people and technology) designed to enhance that process is essentially facilitating innovation. Many people who flaunt innovation cannot even define what it is. Likewise, those who are expected to embrace innovation have no idea what it is. So, a basic discussion of the various views and definitions of innovation is essential. Below are some definitions:

- Innovation is the creation of an enabling environment that facilitates taking advantage of the capabilities of people, process, technology, and other organizational assets.
- Innovation is a new way of doing something.
- Innovation is the methodology of managing, allocating, and timing organizational technology tools, workforce assets, and work processes to achieve a given output in an efficient and expedient manner.
- Innovation is creatively reengineering solutions, for problems which may not yet exist, by actualizing new ideas into valuable processes, services, or goods.
- Innovation is achieving a goal using novel means.
- The process of innovation is the management of resources to drive a new or unexpected result.
- Innovation is a collection of principles to help facilitate problem-solving, looking at both traditional as well as nontraditional approaches.
- Innovation is the project management process of employing novel resources and/or methods to produce a product or a service that is more useful to a current or novel use case.

Organizations will often have their own internal definitions, views, policies, strategies, and guidelines for their own specific operational environments and needs. Of course, in this process, innovation will mean different things to different people based on the specific needs of the individuals. Those individualized views and perceptions will determine how each person reacts to the word "innovation." This then brings about the implication of Maslow's Hierarchy of Needs (Maslow, 1943) in the principles of innovation.

People's Hierarchy of Needs in Embracing Innovation

No innovation effort can be successful without taking care of the people aspects that drive an organization toward its goals. The mental or cognitive capabilities of each person will determine how he or she responds to collaborative opportunities in any innovation environment. A manifestation of Maslow's Hierarchy of Needs can guide each person's responsiveness and adaptation to innovation opportunities. A team member at the lowest level of the hierarchy will respond differently from someone at a higher level of hierarchy of needs. Concurrent with the individual hierarchy of needs is the organizational hierarchy of needs, which is where innovation pursuits normally originate. Figure 1.1 illustrates a multidimensional representation of the dual hierarchy

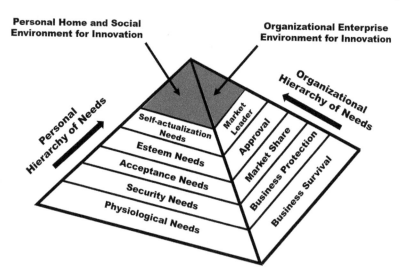

FIGURE 1.1 Individual hierarchy of needs within organizational needs for innovation.

of needs. This forces a consideration of a wider scope of human and organizational needs in the pursuit of innovation.

Communication at each level of hierarchy must be customized for that level. The Triple C principle of project communication (Badiru, 2008) provides the framework for innovation communication, cooperation, and coordination.

Embedded across the alternate definitions are elements of importance to innovation proponents, advocates, sponsors, supporters, sponsors, or observers regarding the questions of what impedes innovation, what facilitates innovation, and how to drive innovation, as summarized below.

Impedance to Innovation

1. What are impediments to innovation?
   a. People's current hierarchy of needs
   b. Risk aversion
   c. Organizational momentum
   d. Stove-piping
   e. Customer isolation
   f. Lack of sufficient time and poor time management
   g. Lack of stakeholder involvement
   h. Lack of structure communication
   i. Lack of innovative practices and understanding built into organizational culture
   j. Excessive risk aversion; fear of failure
   k. Lack of creativity
   l. Lack of diversity of thought
   m. Lack of funds
   n. Burdening regulations
   o. Immovable bureaucracy (e.g., restrictive sources of funding)
   p. Status quo culture
   q. Ignorance about users' satisfaction with current products or services
   r. Fear of technological deviation
   s. Too busy to innovate
   t. Lack of "top-cover" from leadership
   u. Poor collaborating environment
   v. Insufficient resources (tools, budget, personnel)
   w. Organizational quagmire
   x. Customer isolation
   y. Fear of stepping out of the norm
   z. Prior culture of poor cooperation and/or communication.

2. What are facilitators of innovation?
   a. Communicate, communicate, and communicate again
   b. Top-down support and removal of "red tape"
   c. Integrate teams across organization, education, and training
   d. Talk with the customer
   e. Be willing to walk away
   f. Learn from failure
   g. Set SMART goals (Specific, Measurable, Aligned, Realistic/ Achievable, and Timed)
   h. Hold people accountable for established goals
   i. Make innovation a priority
   j. Define risk boundaries
   k. Solicit feedback/listen
   l. Apply lessons learned
   m. Endorse training and promote teamwork
   n. Reward accomplishments; incentivize
   o. Top-down involvement (walk the walk, not just talk the talk)
   p. Get buy in at the lowest levels
   q. Hire diversity to promote diversity of thought
   r. Create space and time for the pursuit of innovation, time to think
   s. Motivate employees with a clear vision
   t. Leadership employing change management
   u. Psychological safety to allow failure
   v. Failure learning via reporting/analysis/synthesis
   w. Create close interaction with various potential users and customers
   x. Institute problem-solving process
   y. Study the elements of innovation within the organization
   z. Embrace novel ideas, pursue, and reward.

INNOVATION DISCIPLES

Who are the disciples of innovation? There are those who embrace, support, fund, advance, and promote innovation. These are the real disciples of innovation, and they should be recognized and supported in their innovation endeavors. The discipleship is demonstrated in a variety of ways. Some write about innovation. Some provide financial support to instigate and facilitate innovation. Some attend innovation events and regurgitate the lessons learned, through which others may become inspired about innovation. Unfortunately, there are those who are innovation naysayers who don't appreciate the benefits of innovation. They are not to be blamed. Their perception is often based on the fact that innovation could be an ambiguous word. The question is if and how innovation is different from the previous waves of total quality management, lean operations, Six Sigma, technology management, artificial intelligence, and so on. A clarification can be helpful to convince the naysayers to come onto the positive side of innovation because, at the end of the day, whatever the prevailing buzzword is, the same goal of achieving goals and objectives promptly, efficiently, and effectively is the underlying push for the buzzword. The end justifies the means. If innovation or any other word gets us moving toward the end goals, then the pathway is comforting and reassuring.

INNOVATION FOR NATIONAL DEFENSE

Since the dawn of history, nations have engrossed themselves in developing new tools, techniques, and methodologies to protect their geographical boundaries. From the crude implements used by prehistorical people to the very modern technologies, the end game has been the same; that is, to protect the homeland. Even in times of peace, efforts must be made to develop new machinery, equipment, processes, and devices targeted for the protection of the nation. The emergence of organized nations and structured communities facilitated even more innovative techniques of national defense. In fact, the drive to achieve new national defense tools led to better underlying science and technology, which most often end up having other beneficial societal applications that are outside the needs of national defense. It is through the efforts of industry that those societal benefits are manifested as practical consumer products. It is important to recognize, document, and demonstrate the paths of converting defense science and technology developments into general industrial applications through deliberate transfer actions. This chapter presents an introductory coverage of the nuances of innovation for national defense.

Innovation is the lifeline of national development. This handbook presents a collection of chapters that provide techniques and methodologies for achieving the transfer of defense-targeted science and technology development for general industrial applications. Experts from national defense institutions, government laboratories, business, and industry contribute chapters to the handbook. The handbook provides a lasting guidance for nations, communities, and businesses expecting to embark upon science and technology transfer to industry under the auspices of national defense pursuits. We don't often make a connection between a viable industrial base and a robust national defense. The fact is that a vibrant base of industrial activities can promote and protect national defense pursuits, particularly where economic vitality is concerned.

There is a need for a good utility framework for this handbook because of the globalization of modern industries desirous of capitalizing on technical developments in the defense industry for the purpose of developing new consumer products. Many nations are interested in embarking on rapid prototyping of new technologies from their defense organizations for the advancement of their nations. Guidelines, strategies, and techniques are needed to actualize their aspirations. Allied nations often conduct joint defense exercises, the coalition from which they can advance their respective local industries. Some good examples of how national defense products enhance general consumer products include the following:

1. There are several consumer products that originated from initial defense focus, such as microwave oven and Global Positioning System (GPS).
2. R&D personnel from defense organizations often end up working in general business and industry, where their expertise is needed through consumer technology transfer processes.
3. The international space station combines the efforts of cooperating nations, thus paving the way for potential advancement of tech-transfer industries at the national level.
4. Many formerly classified defense-related developments have been declassified, thereby necessitating the need for tech-transfer strategies to industry.

The overall conclusion is that a strong national defense program fuels a strong industrial base. Every country, even the poorest ones, must be engaged in national defense pursuits, which are predicated on innovation, both soft and hard. Not all innovation is of a technical breed. Soft innovation may pertain only to the processes and managerial principles for managing and deploying innovation. Hard innovation may relate to technical and technology-based developments that enhance the focus on national defense.

Digital Revolution and Innovation

The digital environment has created new opportunity for new innovation developments both in technology and in operational processes. For example, in the digital emergence of 3D Printing (Additive Manufacturing), the lead editor offers the following operational quotes:

Little thoughts make up big ideas.

Adedeji Badiru

Big components are made in little layers of material.

Adedeji Badiru

Manufacturing is rapidly shifting from manual labor to digital labor through the application of innovation. The digital revolution has landed on the doorstep of conventional manufacturing. What was once limited to the realm of laboratory research has now been transformed, through innovation, to the platform of practicality and reality. For decades, manufacturing had languished within the same old framework of mold-and-cast type of product development. This traditional approach has made manufacturing subject to the inability to respond quickly and adaptively to new product requirements. With the advent of direct digital manufacturing (aka 3D Printing or Additive Manufacturing), product designers and developers now have a mechanism to respond to the requirements for new intricately designed and delivered products, often at the immediate point of need. The manufacturing sector is well positioned to leverage the capabilities of this new digital innovation for designing and making products. The emerging proliferation of 3D Printing in business and industry has made it imperative that a structural forum be organized to guide the path of full utilization of innovative developments in digital manufacturing. The conventional product development environment is vastly different from what 3D Printing will require. Hitherto, individuals and organizations have been jumping on the 3D Printing bandwagon without strategic consideration of downstream and upstream aspects of "printed" products. This book offers a structured platform of enabling innovation in all sectors. Both technical and management issues related to this new wave of innovation are addressed in the book. The expected benefits of innovation dialogue and exchanges include a better alignment of product technology with future developments and the need to secure, maintain, and advance organizational performance. Specifically, readers will learn about the systems engineering aspect of 3D Printing to achieve a faster translation of innovation into real products as well as operational effectiveness, raw material efficiency, higher return on manufacturing investment, rapid and focused product deployment, technology transfer potentials, manufacturing flexibility, and anywhere-anytime agility for product generation.

With the additional emergence of virtual reality (VR), augmented reality (AR), and mixed reality (MR), the platform of innovation for industry is growing rapidly. These emerging technologies can be leveraged to provide cost-effective development of new products. The best way to accomplish this is to mix innovation and collaboration.

Central Role of Innovation

As the military goes, so goes the nation and business climate. The innovation drive by the US military can have a telling effect on how business and industry respond to innovation. The central role of innovation in national defense is evidenced by the fact that "Drive Innovation" is one of the top five priorities announced by the US Air Force in August 2017. The priorities, released by USAF Secretary Heather Wilson, are as follows:

1. Restore readiness
2. Cost-effectively modernize
3. Drive innovation
4. Develop exceptional leaders
5. Strengthen alliances.

We cannot restore readiness without employing new innovative tools and techniques. We cannot cost-effectively modernize without developing and utilizing radically innovative quantitative and qualitative methodologies. We cannot develop exceptional leaders without directing efforts at new, innovative, and specialized education, including advanced education. We cannot strengthen alliances without innovative partnering strategies. In a systems approach, a system is defined as the collection of interrelated elements, whose collective output is higher than the sum of their individual outputs. As a specific tool, the DEJI® (Design, Evaluation, Justification, and Integration) model of systems engineering is unique and innovative because it explicitly calls for a *justification* and *integration* of actions, which requires a more rational decision process during the *design* and *evaluation* stages. The model facilitates a recursive *design-evaluate-justify-integrate* process for enhancing operations. The design stage is essentially the decision stage, which must be evaluated and justified before moving to the implementation stage. The typical implementation stage must be pursued with respect to how well the decision (i.e., design) integrates into the prevailing infrastructure and resource base of the organizations involved. Thus, the model covers the broad spectrum of people, process, and technology in national defense pursuits. Some of the analytical tools used in the DEJI model include state-space modeling, simulation, systems value modeling, learning curve analysis, workload analysis, cognitive modeling, and hierarchical decision transformation. The DEJI model is discussed further later in this chapter. Based on a systems approach, priorities are best pursued from a system of systems perspective. In this regard, multifaceted collaboration approaches must be embraced.

COLLABORATION FOR INNOVATION TRANSFER

Innovation is best pursued via multifaceted collaboration. No one entity has all the answers. Together, innovation is stronger. A framework for academia-government-industry collaboration is essential for leveraging the benefits of innovation for organizational advancement. Just as we may have technology transfer paths, so can we have innovation transfer paths. Innovation transfer is not just about the hardware, technology, or technical components of a system. It can involve a combination of several components, including software (computer-based) and peopleware. Thus, this chapter addresses the transfer of innovation knowledge as well as the transfer of innovation skills.

Due to its many interfaces, the area of technology adoption and implementation is a prime candidate for the application of project planning and control techniques. Technology managers, engineers, and analysts should make an effort to take advantage of the effectiveness of project management tools. This applies the various project management techniques available to the problem of innovation transfer. Project management approach is presented within the context of innovation adoption and implementation for national defense. The Triple C model of communication, cooperation, and coordination is applied as an effective tool for ensuring the acceptance of new innovation product.

To transfer innovation, like any technology transfer, we must know what constitutes innovation. A working definition of innovation will enable us to determine how best to transfer it. A basic question that should be asked is: What is innovation?

Innovation can mean different things to different audiences. Innovation can be defined as follows:

> Innovation is a combination of physical and nonphysical processes that make use of the latest available knowledge, skills, technology, etc. to achieve business, service, or organizational goals.

Innovation is a specialized body of knowledge that can be applied to achieve a mission or purpose. The knowledge concerned could be in the form of methods, processes, techniques, tools, machines, materials, and procedures. Technology design, development, and effective use are driven by effective utilization of human resources and effective management systems. Technological progress is the result obtained when the provision of technology is used in an effective and efficient manner to improve productivity, reduce waste, improve human satisfaction, or meet specific operational needs.

Innovation, all by itself, is useless. However, when the right innovation is put to the right use, with effective supporting management system, it can be very effective in achieving organizational goals. Innovation implementation starts with an idea and ends with a productive process. Innovative progress is said to have occurred when the outputs of innovation, in the form of information, instrument, or knowledge that is used productively and effectively in industrial operations, lead to a lowering of costs of production, better product quality, higher levels of output (from the same amount of inputs), and better alignment with mission requirements. The information and knowledge involved in innovation progress include those that improve the performance of management, labor, and the total resources expended for a given activity.

Innovation progress plays a vital role in improving overall national defense. Experience in the developed countries, such as the United States, shows that in the period of 1870–1957, 90% of the rise in real output per man-hour can be attributed to technological progress fueled by innovation. It is conceivable that a higher proportion of increases in per capita income is accounted for by technological change. Changes occur through improvements in the efficiency in the use of existing technology, that is, through learning and through the adaptation of other technologies, some of which may involve different collections of technological equipment. The challenge to developing countries is how to develop the infrastructure that promotes, uses, adapts, and advances technological knowledge.

Most of the developing nations today face serious challenges arising not only from the worldwide imbalance of dwindling revenue from industrial products and oil, but also from major changes in a world economy that is characterized by competition, imports and exports of not only oil, but also basic technology, weapon systems, and electronics. If technology utilization is not given the right attention in all sectors of the national economy, the much-desired national defense cannot occur or cannot be sustained. If innovation is stymied, the ability of a nation to compete in the world market will, consequently, be stymied, with potential adverse implication for national defense.

The important characteristics or attributes of a new technology innovation may include productivity improvement, improved quality, cost savings, flexibility, reliability, and safety. An integrated evaluation must be performed to ensure that a proposed technology is justified both economically and technically. The scope and goals of the proposed technology must be established right from the beginning of the project. Below is a summary of the common "ilities" related to innovation transfer:

Adaptability: Can the technology be adapted to fit the needs of the organization? Can the organization adapt to the requirements of the technology?

Affordability: Can the organization afford the technology in terms of first-cost, installation cost, sustainment cost, and other incidentals?

Capability: What are the capabilities of the technology with respect to what the organization needs? Can the technology meet the current and emerging needs of the organization?

Compatibility: Is the technology compatible with existing software and hardware?

Configurability: Can the technology be configured for the existing physical infrastructure available within the organization?

Dependability: Is the technology dependable enough to produce the outputs expected?

Desirability: Is the particular technology desirable for the prevailing operating environment of the organization? Are there environmental issues and/or social concerns related to the technology?

Expandability: Can the technology be expanded to fit the changing needs of the organization?

Flexibility: Does the technology have flexible characteristics to accomplish alternate production requirements?

Interchangeability: Can the technology be interchanged with currently available tools and equipment in the organization? In case of operational problems, can the technology be interchanged with something else?

Maintainability: Does the organization have the wherewithal to maintain the technology?

Manageability: Does the organization have adequate management infrastructure to acquire and use the technology?

Reconfigurability: When operating conditions or organizational infrastructure changes, can the technology be reconfigured to meet new needs?

Reliability: Is the technology reliable in terms of technical, physical, and/or scientific characteristics?

Stability: Is the technology mature and stable enough to warrant an investment within the current operating scenario?

Sustainability: Is the organization committed enough to sustain the technology for the long haul? Is the design of the technology sound and proven to be sustainable?

Volatility: Is the technology devoid of volatile developments? Is the source of the technology devoid of political upheavals and/or social unrests?

An assessment of a technology transfer opportunity will entail a comparison of unit-level objectives with the overall organizational goals in the following areas:

1. Marketing and outreach strategy: This should identify the customers of the proposed technology. It should also address items such as market cost of proposed product, assessment of competition, and market share. Import and export considerations should be a key component of the marketing strategy.
2. Industry growth and long-range expectations: This should address short-range expectations, long-range expectations, future competitiveness, future capability, and prevailing size and strength of the industry that will use the proposed technology.
3. National defense benefit: Any prospective technology must be evaluated in terms of direct and indirect benefits to be generated by the technology. These may include product price versus value, increase in international trade, improved standard of living, cleaner environment, safer workplace, and higher productivity.
4. Economic feasibility: An analysis of how the technology will contribute to profitability should consider past performance of the technology, incremental benefits of the new technology versus conventional technology, and value added by the new technology.
5. Capital investment: Comprehensive economic analysis should play a significant role in the technology assessment process. This may cover an evaluation of fixed and sunk costs, cost of obsolescence, maintenance requirements, recurring costs, installation cost, space requirement cost, capital substitution options, return on investment, tax implications, cost of capital, and other concurrent projects.
6. Innovation resource requirements: The utilization of resources (human resources and equipment) in the pre- and post-technology phases of industrialization should be assessed. This may be based on material input–output flows, high value of equipment versus productivity improvement, required inputs for the technology, expected output of the technology, and utilization of technical and nontechnical personnel.
7. Innovation technology stability: Uncertainty is a reality in technology adoption efforts. Uncertainty will need to be assessed for the initial investment, return on investment, payback period, public reactions, environmental impact, and volatility of the technology.
8. National defense improvement: An analysis of how the technology may contribute to national productivity may be verified by studying industrial throughput, efficiency of production processes, utilization of raw materials, equipment maintenance, absenteeism, learning rate, and design-to-production cycle.

Embracing New Innovation

Opportunity lost can be a recurring risk in industry. When new innovation knocks, it should be embraced. A good case example of opportunity lost and innovation ignored is the case of digital photography first developed (and ignored) at Kodak in the mid-1970s. Kodak ignored the new innovation, perhaps because it conflicted with their traditional market model. In 1998, Kodak had 170,000 employees and sold 85% of all photo paper worldwide. Within just a few years, Kodak's business model disappeared and the company went out of its traditional business. Had Kodak aggressively embraced and leveraged the new digital photography in 1975, the future of the company might have taken a different positive and profitable path. If innovation is not promptly embraced and capitalized on, what happened at Kodak can happen to any other companies in the prevailing digital engineering and manufacturing environment, particularly those dealing with artificial intelligence, health, autonomous and electric cars, STEM (science, technology, engineering, and mathematics) education, 3D Printing, agriculture, and knowledge-based jobs.

Fortunately, new industrial and service technologies have been gaining more attention in recent years. This is due to the high rate at which new productivity improvement technologies are being developed. The fast pace of new technologies has created difficult implementation and management problems for many organizations. New technology can be successfully implemented only if it is viewed as a system whose various components must be evaluated within an integrated managerial framework. Such a framework is provided by a project management approach. A multitude of new technologies have emerged in recent years. It is important to consider the peculiar characteristics of a new technology before establishing adoption and implementation strategies. The justification for the adoption of a new technology is usually a combination of several factors rather than a single characteristic of the technology. The potential of a specific technology to contribute to industrial development goals must be carefully assessed. The technology assessment process should explicitly address the following questions:

What is expected from the new technology?
Where and when will the new technology be used?
How is the new technology similar to or different from existing technologies?
What is the availability of technical personnel to support the new technology?
What administrative support is needed for the new technology?

Who will use the new technology?
How will the new technology be used?
Why is the technology needed?

The development, transfer, adoption, utilization, and management of technology is a problem that is faced in one form or another by business, industry, and government establishments. Some of the specific problems in technology transfer and management include the following:

- Controlling technological change
- Integrating technology objectives
- Shortening the technology transfer time
- Identifying a suitable target for technology transfer
- Coordinating the research and implementation interface
- Formal assessment of current and proposed technologies
- Developing accurate performance measures for technology
- Determining the scope or boundary of technology transfer
- Managing the process of entering or exiting a technology
- Understanding the specific capability of a chosen technology
- Estimating the risk and capital requirements of a technology.

Integrated managerial efforts should be directed at the solution of the problems stated above. A managerial revolution is needed in order to cope with the ongoing technological revolution. The revolution can be initiated by modernizing the long-standing and obsolete management culture relating to technology transfer. Some of the managerial functions that will need to be addressed when developing a technology transfer strategy include the following:

1. Development of an innovation and technology transfer plan.
2. Assessment of technological risk.
3. Assignment/reassignment of personnel to implement the technology transfer.
4. Establishment of a transfer manager and a technology transfer office. In many cases, transfer failures occur because no individual has been given the responsibility to ensure the success of technology transfer.
5. Identification and allocation of the resources required for technology transfer.
6. Setting of guidelines for technology transfer. For example,
   a. Specification of phases (development, testing, transfer, etc.)
   b. Specification of requirements for interphase coordination

    c. Identification of training requirements.
    d. Establishment and implementation of performance measurement.
7. Identify key factors (both qualitative and quantitative) associated with technology transfer and management.
8. Investigate how the factors interact and develop the hierarchy of importance for the factors.
9. Formulate a loop system model that considers the forward and backward chains of actions needed to effectively transfer and manage a given technology.
10. Track the outcome of the technology transfer.

Technological developments in many industries appear in scattered, narrow, and isolated areas within a few selected fields. This makes technology efforts to be rarely coordinated, thereby hampering the benefits of technology. The optimization of technology utilization is, thus, very difficult. To overcome this problem and establish the basis for effective technology transfer and management, an integrated approach must be followed. An integrated approach will be applicable to technology or innovation transfer between any two organizations, whether public or private.

Some nations concentrate on the acquisition of bigger, better, and faster technology. But little attention is given to how to manage and coordinate the operations of the technology once it arrives. When technology fails, it is not necessarily because the technology is deficient. Rather, it is often the communication, cooperation, and coordination functions of technology management that are deficient. Technology encompasses factors and attributes beyond mere hardware, software, and peopleware, which refers to people issues affecting the utilization of technology. This may involve socioeconomic and cultural issues of using certain technologies or innovative techniques. Consequently, innovation transfer involves more than the physical transfer of hardware and software. Several flaws exist in the common practices of technology transfer and management. These flaws include the following:

- Poor fit: This relates to an inadequate assessment of the need of the organization receiving the technology. The target of the transfer may not have the capability to properly absorb the technology.
- Premature transfer of technology: This is particularly acute for emerging technologies that are prone to frequent developmental changes.
- Lack of focus: In the attempt to get a bigger share of the market or gain early lead in the technological race, organizations frequently force technology in many incompatible directions.

- Intractable implementation problems: Once a new technology is in place, it may be difficult to locate sources of problems that have their roots in the technology transfer phase itself.
- Lack of transfer precedents: Very few precedents are available on the management of brand new technology. Managers are, thus, often unprepared for their new technology management responsibilities.
- Stuck on technology: Unworkable technologies sometimes continue to be recycled needlessly in the attempt to find the "right" usage.
- Lack of foresight: Due to the nonexistence of a technology transfer model, managers may not have a basis against which they can evaluate future expectations.
- Insensitivity to external events: Some external events that may affect the success of technology transfer may include trade barriers, taxes, and political changes.
- Improper allocation of resources: There is usually not enough resources available to allocate to technology alternatives. Thus, a technology transfer priority must be developed.

The following steps provide a specific guideline for pursuing the implementation of manufacturing technology transfer:

1. Find a suitable application.
2. Commit to an appropriate technology.
3. Perform economic justification.
4. Secure management support for the chosen technology.
5. Design the technology implementation to be compatible with existing operations.
6. Formulate project management approach to be used.
7. Prepare the receiving organization for the technology change.
8. Install the technology.
9. Maintain the technology.
10. Periodically review the performance of the technology based on prevailing goals.

Innovation Transfer Modes

The transfer of technology can be achieved in various forms. Project management provides an effective means of ensuring proper transfer of technology. Three technology transfer modes are presented here to illustrate basic strategies for getting one technological product from one point (technology source) to another (technology sink). The elements of innovation transfer can encompass the following elements:

Operating concepts
Operating process
Physical products
Technical hardware
Software tool
Workforce.

In a feedforward–feedback relationship, at the other end of the innovation reverse-transfer spectrum, the pathway may include the following:

- Identical innovation
- Totally new innovation
- Totally new process.

Innovation and technology application centers may be established to serve as a unified point for linking technology sources with interested targets. The center will facilitate interactions between business establishments, academic institutions, and government agencies to identify important technology needs. Technology can be transferred in one or a combination of the following strategies:

1. Transfer of complete technological products: In this case, a fully developed product is transferred from a source to a target. Very little product development effort is carried out at the receiving point. However, information about the operations of the product is fed back to the source so that necessary product enhancements can be pursued. So, the technology recipient generates product information which facilitates further improvement at the technology source. This is the easiest mode of technology transfer and the most tempting. Developing nations are particularly prone to this type of transfer. Care must be exercised to ensure that this type of technology transfer does not degenerate into "machine transfer." It should be recognized that machines alone do not constitute technology.
2. Transfer of technology procedures and guidelines: In this technology transfer mode, procedures (e.g., Blueprints) and guidelines are transferred from a source to a target. The technology blueprints are implemented locally to generate the desired services and products. The use of local raw materials and manpower is encouraged for the local production. Under this mode, the implementation of

the transferred technology procedures can generate new operating procedures that can be fed back to enhance the original technology. With this symbiotic arrangement, a loop system is created whereby both the transferring and the receiving organizations derive useful benefits.

3. Transfer of technology concepts, theories, and ideas: This strategy involves the transfer of the basic concepts, theories, and ideas behind a given technology. The transferred elements can then be enhanced, modified, or customized within local constraints to generate new technological products. The local modifications and enhancements have the potential to generate an identical technology, a new related technology, or a new set of technology concepts, theories, and ideas. These derived products may then be transferred back to the original technology source as new technological enhancements. In a source–sink linkage, a specific cycle for local adaptation and modification of technology may be developed. An academic institution is a good potential source for the transfer of technology concepts, theories, and ideas. Thus, the innovation transfer loop links elements from the source to elements in the sink (target) as a form of innovation export.

It is very important to determine the mode in which technology will be transferred for defense purposes. There must be a concerted effort by people to make the transferred technology work within local infrastructure and constraints. Local innovation, patriotism, dedication, and willingness to adapt technology will be required to make technology transfer successful. It will be difficult for a nation to achieve national defense through total dependence on transplanted technology. Local adaptation will always be necessary.

CHANGEOVER STRATEGIES FOR INNOVATION

One good innovation begets another. Thus, changeover arrangements are essential for a smooth transition between stages of innovation. Any development project will require changing from one form of technology to another. The implementation of a new technology to replace an existing (or a nonexistent) technology can be approached through one of several options. Some options are more suitable than others for certain types of technologies. The most commonly used technology changeover strategies include the following:

Parallel changeover: In this case, the existing technology and the new technology operate concurrently until there is confidence that the new technology is satisfactory.

Direct changeover: In this approach, the old technology is removed totally and the new technology takes over. This method is recommended only when there is no existing technology or when both technologies cannot be kept operational due to incompatibility or cost considerations.

Phased changeover: In this incremental changeover method, modules of the new technology are gradually introduced one at a time using either direct or parallel changeover.

Pilot changeover: In this case, the new technology is fully implemented on a pilot basis in a selected department within the organization.

Post-Implementation Evaluation of Innovation

The new technology should be evaluated only after it has reached a steady-state performance level. This helps to avoid the bias that may be present at the transient stage due to personnel anxiety, lack of experience, or resistance to change. The system should be evaluated for the following aspects:

- Sensitivity to data errors
- Quality and productivity
- Utilization level
- Response time
- Effectiveness.

Innovation Systems Integration

With the increasing shortages of resources, more emphasis should be placed on the sharing of resources. Technology resource sharing can involve physical equipment, facilities, technical information, ideas, and related items. The integration of technologies facilitates the sharing of resources. Technology integration is a major effort in technology adoption and implementation. Technology integration is required for proper product coordination. Integration facilitates the coordination of diverse technical and managerial efforts to enhance organizational functions, reduce cost, improve productivity, and increase the utilization of resources. Technology integration ensures that all performance goals are satisfied with a minimum of expenditure of time and resources. It may require the adjustment of functions to permit sharing of resources, development of new policies to accommodate product integration, or realignment of managerial responsibilities. It can affect both hardware and software components of an organization. Important factors in technology integration include the following:

- Unique characteristics of each component in the integrated technologies
- Relative priorities of each component in the integrated technologies
- How the components complement one another
- Physical and data interfaces between the components
- Internal and external factors that may influence the integrated technologies
- How the performance of the integrated system will be measured.

Government-Driven Innovation Transfer

The malignant policies and operating characteristics of some of the governments in underdeveloped countries have contributed to stunted growth of technology in those parts of the world. The governments in most developing countries control the industrial and public sectors of the economy. Either people work for the government or serve as agents or contractors for the government. The few industrial firms that are privately owned depend on government contracts to survive. Consequently, the nature of the government can directly determine the nature of industrial technological progress.

The operating characteristics of most of the governments perpetuate inefficiency, corruption, and bureaucratic bungles. This has led to a decline in labor and capital productivity in the industrial sectors. Using the Pareto distribution, it can be estimated that in most government-operated companies, there are eight administrative workers for every two production workers. This creates a nonproductive environment that is skewed toward hyper-bureaucracy. The government of a nation pursuing industrial development must formulate and maintain an economic stabilization policy. The objective should be to minimize the sacrifice of economic growth in the short run while maximizing long-term economic growth. To support industrial technology transfer efforts, it is essential that a conducive national policy be developed.

More emphasis should be placed on industry diversification, training of the work force, supporting financial structure for emerging firms, and implementing policies that encourage productivity in a competitive economic environment. Appropriate foreign exchange allocation, tax exemptions, bank loans for emerging businesses, and government-guaranteed low-interest loans for potential industrial entrepreneurs are some of the favorable policies to spur growth and development of the industrial sector.

Improper trade and domestic policies have adversely affected industrialization in many countries. Excessive regulations that cause bottlenecks in industrial enterprises are not uncommon. The regulations can take the form of licensing, safety requirements, manufacturing value-added quota requirements, capital contribution by multinational firms, and high domestic production protection. Although regulations are needed for industrial operations, excessive controls lead to low returns from the industrial sectors. For example, stringent regulations on foreign exchange allocation and control have led to the closure of industrial plants in some countries. The firms that cannot acquire essential raw materials, commodities, tools, equipment, and new technology from abroad due to foreign exchange restrictions are forced to close and lay off workers.

Price controls for commodities are used very often by developing countries especially when inflation rates for essential items are high. The disadvantages involved in price control of industrial goods include restrictions of the free competitive power of available goods in relation to demand and supply, encouragement of inefficiency, promotion of dual markets, distortion of cost relationships, and increase in administrative costs involved in producing goods and services.

One way that a government can help facilitate industrial technology transfer involves the establishment of technology transfer centers within appropriate government agencies. A good example of this approach can be seen in the government-sponsored technology transfer program by the US National Aeronautics and Space Administration (NASA). In the Space Act of 1958, the US Congress charged NASA with a responsibility to provide for the widest practical and appropriate dissemination of information concerning its activities and the results achieved from those activities. With this technology transfer responsibility, technology developed in the United State's space program is available for use by the nation's business and industry.

In order to accomplish technology transfer to industry, NASA established a Technology Utilization Program (TUP) in 1962. The TUP uses several avenues to disseminate information on NASA technology. The avenues include the following:

- Complete, clear, and practical documentation is required for new technology developed by NASA and its contractors. These are available to industry through several publications produced by NASA. An example is a monthly, Tech Briefs, which outlines technology innovations. This is a source of prompt technology information for industry.

- Industrial Application Centers (IACs) were developed to serve as repositories for vast computerized data on technical knowledge. The IACs are located at academic institutions around the country. All the centers have access to a large database containing millions of NASA documents. With this database, industry can have access to the latest technological information quickly. The funding for the centers are obtained through joint contributions from several sources including NASA, the sponsoring institutions, and state government subsidies. Thus, the centers can provide their services at very reasonable rates.

- NASA operates a Computer Software Management and Information Center (COSMIC) to disseminate computer programs developed through NASA projects. COSMIC, which is located at a university, has a library of thousands of computer programs. The center publishes an annual index of available software.

In addition to the specific mechanisms discussed above, NASA undertakes application engineering projects. Through these projects, NASA collaborates with industry to modify aerospace technology for use in industrial applications. To manage the application projects, NASA established a Technology Application Team (TAT), consisting of scientists and engineers from several disciplines. The team interacts with NASA field centers, industry, universities, and government agencies. The major mission of the team interactions is to define important technology needs and identify possible solutions within NASA. NASA applications engineering projects are usually developed in a five-phase approach with go or no-go decisions made by NASA and industry at the completion of each phase. The five phases are outlined below:

1. NASA and the TAT meet with industry associations, manufacturers, university researchers, and public sector agencies to identify important technology problems that might be solved by aerospace technology.

2. After a problem is selected, it is documented and distributed to the Technology Utilization Officer at each of NASA's field centers. The officer in turn distributes the description of the problem to the appropriate scientists and engineers at the center. Potential solutions are forwarded to the team for review. The solutions are then screened by the problem originator to assess the chances for technical and commercial success.

3. The development of partnerships and a project plan to pursue the implementation of the proposed solution. NASA joins forces with private companies and other organizations to develop an applications

engineering project. Industry participation is encouraged through a variety of mechanisms such as simple letters of agreement or joint endeavor contracts. The financial and technical responsibilities of each organization are specified and agreed upon.

4. At this point, NASA's primary role is to provide technical assistance to facilitate utilization of the technology. The costs for these projects are usually shared by NASA and the participating companies. The proprietary information provided by the companies and their rights to new discoveries are protected by NASA.

5. The final phase involves the commercialization of the product. With the success of commercialization, the project would have widespread impact. Usually, the final product development, field testing, and marketing are managed by private companies without further involvement from NASA.

Through this well-coordinated, government-sponsored technology transfer program, NASA has made significant contributions to the US industry, thereby providing an anchor for national defense pursuits. The results of NASA's technology transfer abound in numerous consumer products either in subtle forms or in clearly identifiable forms. Food preservation techniques constitute one area of NASA's technology transfer that has had a significant positive impact on the society. Although the specific organization and operation of the NASA technology transfer programs have changed in name or in deed over the years, the basic descriptions outlined above remain a viable template for how to facilitate manufacturing technology transfer. In a similar government-backed strategy, the US Air Force Research Lab (AFRL) also has very structured programs for transferring nonclassified technology to the industrial sector. It is believed that a project management approach can help in facilitating success with innovation and technology transfer efforts.

Picking the Right Innovation

It is important to pick the right innovation to adopt and adapt. The question of which innovation is appropriate to transfer in or transfer out is relevant for technology transfer considerations. While several methods of technology selection are available, this book recommends methods that combine qualitative and quantitative factors. The analytical hierarchy process (AHP) is one such method. Another useful, but less publicized is the PICK chart. The PICK chart was originally developed by Lockheed Martin to identify and prioritize improvement opportunities in the company's process improvement applications. The technique is just one of the several decision tools available in process improvement endeavors. It is a very effective technology selection tool used to categorize ideas and opportunities. The purpose is to qualitatively help identify the most useful ideas. A 2×2 grid is normally drawn on a white board or large flip chart. Ideas that were written on sticky notes by team members are placed on the grid based on a group assessment of the payoff relative to the level of difficulty. The PICK acronym comes from the labels for each of the quadrants of the grid: Possible (easy, low payoff), Implement (easy, high payoff), Challenge (hard, high payoff), and Kill (hard, low payoff). The PICK chart quadrants are summarized as follows:

Possible (easy, low payoff)   ➜ Third quadrant
Implement (easy, high payoff) ➜ Second quadrant
Challenge (hard, high payoff) ➜ First quadrant
Kill (hard, low payoff)   ➜ Fourth quadrant

The primary purpose is to help identify the most useful ideas, especially those that can be accomplished immediately with little difficulty. These are called "Just-Do-Its." The general layout of the PICK chart grid is shown in Figure 1.2. The PICK process is normally done subjectively by a team of decision makers under a group decision process. This can lead to bias and protracted debate of where each item belongs. It is desired to improve the efficacy of the process by introducing some quantitative analysis. Badiru and Thomas (2013) present a methodology to achieve a quantification of the PICK selection process.

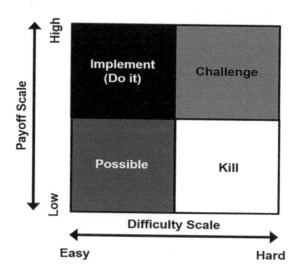

FIGURE 1.2 PICK chart for picking innovation adoption.

The PICK chart is often criticized for its subjective rankings and lack of quantitative analysis. The approach presented by Badiru and Thomas (2013) alleviates such concerns by normalizing and quantifying the process of integrating the subjective rakings by those involved in the group PICK process. Human decision is inherently subjective. All we can do is to develop techniques to mollify the subjective inputs rather than compounding them with subjective summarization.

The placement of items into one of the four categories in a PICK chart is done through expert ratings, which are often subjective and nonquantitative. In order to put some quantitative basis to the PICK chart analysis, Badiru and Thomas (2013) present the methodology of dual numeric scaling on the impact and difficulty axes. Suppose each technology is ranked on a scale of one to ten and plotted accordingly on the PICK chart. Then, each project can be evaluated on a binomial pairing of the respective rating on each scale. Note that a high rating along the x axis is desirable while a high rating along the y axis is not desirable. Thus, a composite rating involving x and y must account for the adverse effect of high values of y. A simple approach is to define $y' = (11-y)$, which is then used in the composite evaluation. If there are more factors involved in the overall project selection scenario, the other factors can take on their own lettered labeling (e.g., a, b, c, z). Then, each project will have an n-factor assessment vector. In its simplest form, this approach will generate a rating such as the following:

$$PICK_{R,i}(x, y') = x + y'$$

where
$PICK_{R,i}(x,y) = $ PICK rating of project i ($i = 1, 2, 3, \ldots, n$)
$n = $ number of project under consideration
$x = $ rating along the impact axis ($1 \leq x \leq 10$)
$y = $ rating along the difficulty axis ($1 \leq y \leq 10$)
$y' = (11 - y)$.

If $x + y'$ is the evaluative basis, then each technology's composite rating will range from 2 to 20, 2 being the minimum and 20 being the maximum possible. If $(x)(y)$ is the evaluative basis, then each project's composite rating will range from 1 to 100. In general, any desired functional form may be adopted for the composite evaluation. Another possible functional form is

$$PICK_{R,i}(x, y'') = f(x, y'')$$

$$= (x + y'')^2,$$

where y'' is defined as needed to account for the converse impact of the axes of difficulty. The above methodology provides a quantitative measure for translating the entries in a conventional PICK chart into an analytical technique to rank the technology alternatives, thereby reducing the level of subjectivity in the final decision. The methodology can be extended to cover cases where a technology has the potential to create negative impacts, which may impede organizational advancement.

The quantification approach facilitates a more rigorous analytical technique compared to traditional subjective approaches. One concern is that although quantifying the placement of alternatives on the PICK chart may improve the granularity of relative locations on the chart, it still does not eliminate the subjectivity of how the alternatives are assigned to quadrants in the first place. This is a recognized feature of many decision tools. This can be mitigated by the use of additional techniques that aid decision makers to refine their choices. The AHP could be useful for this purpose. Quantifying subjectivity is a continuing challenge in decision analysis. The PICK chart quantification methodology offers an improvement over the conventional approach.

Although the PICK chart has been used extensively in industry, there are few published examples in the open literature. The quantification approach presented by Badiru and Thomas (2013) may expand interest and applications of the PICK chart among technology researchers and practitioners. The steps for implementing a PICK chart are summarized below:

Step 1: On a chart, place the subject question. The question needs to be asked and answered by the team at different stages to be sure that the data that is collected is relevant.

Step 2: Put each component of the data on a different note like a post-it or small cards. These notes should be arranged on the left side of the chart.

Step 3: Each team member must read all notes individually and consider its importance. The team member should decide whether the element should or should not remain a fraction of the significant sample. The notes are then removed and moved to the other side of the chart. Now, the data is condensed enough to be processed for a particular purpose by means of tools that allow groups to reach a consensus on priorities of subjective and qualitative data.

Step 4: Apply the quantification methodology presented above to normalize the qualitative inputs of the team.

CONCLUSION

Technology transfer is a great avenue to advancing industrialization. This chapter has presented a variety of principles, tools, techniques, and strategies useful for managing technology transfer. Of particular emphasis in the chapter is the management aspects of innovation technology transfer. The technical characteristics of the technology of interest are often well understood. What is often lacking is an appreciation of the technology management requirements for achieving a successful technology transfer. This chapter presents the management aspects of innovation technology transfer.

REFERENCES

Badiru, A.B. (2008). *Triple C Model of Project Management: Communication, Cooperation, and Coordination.* Boca Raton, FL: CRC Press/Taylor & Francis Group.

Badiru, A.B. and Thomas, M. (2013). Quantification of the PICK chart for process improvement decisions, Journal of Enterprise Transformation, 3(1), 1–15.

Maslow, A.H. (1943). A theory of human motivation, *Psychological Review*, 50(4), 370–396.

Structured Management of Innovation Projects

2

PROJECT SYSTEMS FRAMEWORK FOR INNOVATION

Managing innovation is akin to managing a complex project (Badiru, 2019). The pursuit of innovation is a project that lends itself to project management techniques from a systems perspective. A systems view of innovation makes innovation more agile, efficient, effective, and sustainable. Every innovation effort should be viewed as a project. Thereby, the typical management principles of project management are applicable to the management of innovation pursuits. A system is a collection of interrelated elements working together synergistically to achieve a set of objectives.

Innovation is akin to a project system. Any project is, in actuality, a collection of interrelated activities, people, tools, resources, processes, and other assets brought together in the pursuit of a common goal. The goal may be in terms of generating a physical product, providing a service, or achieving a specific result. This makes it possible to view any project as a system that is amenable to all the classical and modern concepts of systems management. Project systems management is the foundation of everything we do. Having a knowledge is not enough, we must apply the knowledge strategically and systematically for it to be of any use. The knowledge must be applied to do something in the pursuit of objectives. Project management facilitates the application of

31

knowledge and willingness to actually accomplish tasks. Where there is knowledge, willingness often follows. But it is the project execution that actually gets jobs accomplished. From the very basic tasks to the very complex endeavors, project management must be applied to get things done. It is, thus, essential that project management be a part of the core of every academic curriculum in any discipline, whether it is in the liberal arts, medical science, business, retail, education, science, advanced technology, or engineering. The tools and techniques presented in this book are generally applicable to any project-oriented pursuit in business, industry, education, the military, and government. This practically means everything that everyone does because every pursuit can, indeed, be defined as a project. Even a national political process is amenable to a rigorous application of project management tools and techniques. In this regard, a systems approach is of utmost importance in any human pursuit.

Classical control system focuses on control of the dynamics of mechanical objects, such as a pump, electrical motor, turbine, rotating wheel, and so on. The mathematical basis for such control systems can be adapted (albeit in iconic formats) for organizational management systems, including project management. This is because both technical and managerial systems are characterized by inputs, variables, processing, control, feedback, and output. This is represented graphically by input–process–output relationship block diagrams. Mathematically, it can be represented as

$$z = f(x) + \varepsilon$$

where
   z is the output
   $f()$ is the functional relationship
   ε is the error component (noise, disturbance, etc.).

For multivariable cases, as the case of innovation, the mathematical expression is represented as vector–matrix functions as shown in the following:

$$\mathbf{Z} = \mathbf{f(X)} + \mathbf{E}$$

where
   Each term is a matrix
   $\mathbf{Z}$ is the output vector
   $\mathbf{f}(\cdot)$ is the input vector
   $\mathbf{E}$ is the error vector.

Regardless of the level or form of mathematics used, all systems exhibit the same input–process–output characteristics, either quantitatively or qualitatively.

The premise of this book is that there should be a cohesive coupling of quantitative and qualitative approaches in managing a project system. In fact, it is this unique blending of approaches that makes systems application for project management more robust than what one will find in mechanical control systems, where the focus is primarily on quantitative representations.

Organizational performance is predicated on a multitude of factors, some are quantitative while some are qualitative. Systems engineering efficiency and effectiveness are of interest across the spectrum of the diversity of organizational performance under the platform of project management. Project analysts should be interested in having systems engineering serve as the umbrella for improvement efforts throughout the organization. This will get everyone properly connected with the prevailing organizational goals as well as create collaborative avenues among the personnel. Systems application applies across the spectrum of any organization and encompasses the following elements:

Technological systems (e.g., engineering control systems and mechanical systems)

Organizational systems (e.g., work process design and operating structures)

Human systems (e.g., interpersonal relationships and human–machine interfaces).

A systems view of the world makes everything work better and projects more likely to succeed. A systems view provides a disciplined process for the design, development, and execution of complex projects, both in engineering and nonengineering organizations. One of the major advantages of a systems approach is the win–win benefit for everyone. A systems view of innovation also advocates the full involvement of all stakeholders. The Confucius quote below highlights the importance of full involvement:

Tell me and I forget;
Show me and I remember;
Involve me and I understand.

– Confucius, Chinese Philosopher

For example, the pursuit of organizational or enterprise transformation is best achieved through the involvement of everyone, from a systems perspective. Every project environment is very complex because of the diversity of factors involved. There are differing human personalities. There are differing technical requirements. There are differing expectations. There are differing environmental factors. Each specific context and prevailing circumstances determine the specific flavor of what can and cannot be done in the project.

The best approach for effective project management is to adapt to what each project needs. This requires taking a systems view of the project of innovation. This is an essential requirement in today's globalized and intertwined project goals. A systems view requires a disciplined embrace of multidisciplinary execution of projects in a way that components complement each other in the project system. Project management represents an excellent platform for the implementation of a systems approach. Project management integrates various technical and management requirements. It requires control techniques, such as operations research, operations management, forecasting, quality control, and simulation to deliver goals. Traditional approaches to project management use these techniques in a disjointed fashion, thus ignoring the potential interplay among the techniques. One definition of systems project management offered here is stated as follows:

> Innovation systems project management is the process of using systems approach to manage, allocate, and time resources to achieve systems-wide innovation goals in an efficient and expeditious manner.

The definition calls for a systematic integration of technology, human resources, and work process design to achieve goals and objectives. There should be a balance in the synergistic integration of humans and technology. There should not be an overreliance on technology, nor should there be an overdependence on human processes. Similarly, there should not be too much emphasis on analytical models to the detriment of commonsense human-based decisions.

Systems engineering is growing in appeal as an avenue to achieve organizational goals and improve operational effectiveness and efficiency. Researchers and practitioners in business, industry, and government are all clamoring collaboratively for systems engineering implementations. So, what is systems engineering? Several definitions exist. The following is one quite comprehensive definition:

> Systems engineering is the application of engineering to solutions of a multifaceted problem through a systematic collection and integration of parts of the problem with respect to the life cycle of the problem. It is the branch of engineering concerned with the development, implementation, and use of large or complex systems. It focuses on specific goals of a system considering the specifications, prevailing constraints, expected services, possible behaviors, and structure of the system. It also involves a consideration of the activities required to assure that the system's performance matches the stated goals.

Systems engineering addresses the integration of tools, people, and processes required to achieve a cost-effective and timely operation of the system.

INNOVATION LOGISTICS

Logistics can be defined as the planning and implementation of a complex task, the planning and control of the flow of goods and materials through an organization or manufacturing process, or the planning and organization of the movement of personnel, equipment, and supplies. Complex projects represent a hierarchical system of operations. Thus, we can view a project system as collection of interrelated projects all serving a common end goal. Consequently, we present the following universal definition:

> Project systems logistics is the planning, implementation, movement, scheduling, and control of people, equipment, goods, materials, and supplies across the interfacing boundaries of several related projects.
> Conventional project management must be modified and expanded to address the unique logistics of project systems.

INNOVATION SYSTEMS VALUE AND CONSTRAINTS

Systems management is the pursuit of organizational goals within the constraints of time, cost, and quality expectations. The iron triangle model shows that project accomplishments are constrained by the boundaries of quality, time, and cost. In this case, quality represents the composite collection of project requirements. In a situation where precise optimization is not possible, there will have to be trade-offs between these three factors of success. The concept of iron triangle is that a rigid triangle of constraints encases the project. Everything must be accomplished within the boundaries of time, cost, and quality. If better quality is expected, a compromise along the axes of time and cost must be executed, thereby altering the shape of the triangle.

The trade-off relationships are not linear and must be visualized in a multidimensional context. This is better articulated by a three-dimensional view of the systems constraints of the iron triangle. Scope requirements determine the project boundary, and trade-offs must be done within that boundary.

Systems-influence philosophy suggests the realization that you control the internal environment while only influencing the external environment. A "blobby" environment is characterized by intractable activities where everyone is busy, but without a cohesive structure of input–output relationships. In such a case, the following disadvantages may be present:

Lack of traceability
Lack of process control
Higher operating cost
Inefficient personnel interfaces
Unrealized technology potentials.

Organizations often inadvertently fall into the blobs structure because it is simple, low cost, and less time consuming until a problem develops. A desired alternative is to model the project system using a systems value stream structure. This uses a proactive and problem-preempting approach to execute projects. This alternative has the following advantages:

Problem diagnosis is easier
Accountability is higher
Operating waste is minimized
Conflict resolution is faster
Value points are traceable.

A technique that can be used to assess overall value-added components of a process improvement program is the systems value model (SVM), which is an adaptation of the manufacturing system value (MSV) model presented by Troxler and Blank (1989). The model provides an analytical decision aid for comparing process alternatives. Value is represented as a p-dimensional vector:

$$V = f\left(A_1, A_2, \ldots, A_p\right)$$

where $A = (A_1, \ldots, A_n)$ is a vector of quantitative measures of tangible and intangible attributes. Examples of process attributes are quality, throughput, capability, productivity, cost, and schedule. Attributes are considered to be a combined function of factors, x_1, expressed as

$$A_k\left(x_1, x_2, \ldots, x_{m_k}\right) = \sum_{i=1}^{m_k} f_i\left(x_i\right)$$

where

$\{x_i\}$ is the set of m factors associated with attribute A_k ($k = 1, 2, \ldots, p$)
f_i is the contribution function of factor x_i to attribute A_k.

Examples of factors include reliability, flexibility, user acceptance, capacity utilization, safety, and design functionality. Factors are themselves considered to be composed of indicators, v_i, expressed as

$$x_i\left(v_1, v_2, \ldots, v_n\right) = \sum_{j=1}^{n} z_i\left(v_i\right)$$

where

$\{v_j\}$ is the set of n indicators associated with factor x_i ($i = 1, 2, \ldots, m$)
z_j is the scaling function for each indicator variable v_j.

Examples of indicators are project responsiveness, lead time, learning curve, and work rejects. By combining the aforementioned definitions, a composite measure of the value of a process can be modeled mathematically.

A subjective measure to indicate the utility of the decision maker may be included in the model by using an attribute weighting factor, w_i, to obtain a weighted PV:

$$PV_w = f\left(w_1 A_1, w_2 A_2, \ldots, w_p A_p\right)$$

where

$$\sum_{k=1}^{p} w_k = 1 \quad \left(0 \le w_k \le 1\right)$$

With this modeling approach, a set of process options can be compared on the basis of a set of attributes and factors.

Project management should be an enterprise-wide, systems-based endeavor. Enterprise-wide project management is the application of project management techniques and practices across the full scope of the enterprise. This concept is also referred to as management by project (MBP). MBP is a contemporary concept that employs project management techniques in various functions within an organization. MBP recommends pursuing endeavors as project-oriented activities. It is an effective way to conduct any business activity. It represents a disciplined approach that defines any work assignment as a project. Under MBP, every undertaking is viewed as a project that must be managed just like a traditional project. The characteristics required of each project so defined are

1. An identified scope and a goal
2. A desired completion time
3. Availability of resources
4. A defined performance measure
5. A measurement scale for review of work.

An MBP approach to operations helps in identifying unique entities within functional requirements. This identification helps determine where functions overlap and how they are interrelated, thus paving the way for better planning, scheduling, and control. Enterprise-wide project management facilitates a unified view of organizational goals and provides a way for project teams to use information generated by other departments to carry out their functions.

The use of project management continues to grow rapidly. The need to develop effective management tools increases with the increasing complexity of new technologies and processes. The life cycle of a new product to be introduced into a competitive market is a good example of a complex process that must be managed with integrative project management approaches. The product will encounter management functions as it goes from one stage to the next. Project management will be needed throughout the design and production stages of the product. Project management will be needed in developing marketing, transportation, and delivery strategies for the product. When the product finally gets to the customer, project management will be needed to integrate its use with those of other products within the customer's organization.

The need for a project management approach is established by the fact that a project will always tend to increase in size even if its scope is narrowing. The following three literary laws are applicable to any project environment:

Parkinson's law: Work expands to fill the available time or space.
Peter's principle: People rise to the level of their incompetence.
Murphy's law: Whatever can go wrong will.
Badiru's rule: The grass is always greener where you most need it to be dead.

An integrated systems project management approach can help diminish the adverse impacts of these laws in an innovation environment through good communication, project planning, organizing, scheduling, and control.

TOOLS FOR INNOVATION MANAGEMENT

Project management tools can be classified into three major categories:

1. *Qualitative tools:* There are the managerial tools that aid in the interpersonal and organizational processes required for project management.
2. *Quantitative tools:* These are analytical techniques that aid in the computational aspects of project management.
3. *Computer tools:* These are software and hardware tools that simplify the process of planning, organizing, scheduling, and controlling a project. Software tools can help in both the qualitative and quantitative analyses needed for project management.

Although individual books dealing with management principles, optimization models, and computer tools are available, there are few guidelines for the integration of the three areas for project management purposes. In this book, we integrate these three areas for a comprehensive guide to project management. The book introduces the *triad approach* to improve the effectiveness of project management with respect to schedule, cost, and performance constraints within the context of systems modeling. It is one thing to have a quantitative model, but it is a different thing to be able to apply the model to real-world problems in a practical form. The systems approach facilitates making the transition from model to practice.

A systems approach helps increase the intersection of the three categories of project management tools and, hence, improve overall management effectiveness. Crisis should not be the instigator for the use of project management techniques. Project management approaches should be used upfront to prevent avoidable problems rather than to fight them when they develop. What is worth doing is worth doing well, right from the beginning.

CRITICAL FACTORS FOR INNOVATION SUCCESS

The critical factors for innovation success revolve around people and the personal commitment and dedication of each person. No matter how good a technology is and no matter how enhanced a process might be, it is ultimately the people involved that determine success. This makes it imperative to take care of people issues first in the overall systems approach to project management. Many organizations recognize this, but only few have been able to actualize the ideals of managing people productively. Execution of operational strategies requires forthrightness, openness, and commitment to get things done. Lip service and arm waving are not sufficient. Tangible programs that cater to the needs of people must be implemented. It is essential to provide incentives, encouragement, and empowerment for people to be self-actuating in determining how best to accomplish their job functions. A summary of critical factors for systems success encompasses the following:

Total system management (hardware, software, and people)
Operational effectiveness
Operational efficiency
System suitability
System resilience
System affordability
System supportability
System life cycle cost
System performance
System schedule
System cost.

Systems engineering tools, techniques, and processes are essential for project life cycle management to make goals possible within the context of *SMART* principles, which are represented as follows:

1. *Specific:* Pursue specific and explicit outputs.
2. *Measurable:* Design of outputs that can be tracked, measured, and assessed.
3. *Achievable:* Make outputs to be achievable and aligned with organizational goals.
4. *Realistic:* Pursue only the goals that are realistic and result oriented.
5. *Timed:* Make outputs timed to facilitate accountability.

Systems engineering provides the technical foundation for executing innovation successfully. A systems approach is particularly essential in the early stages of the pursuit of innovation in order to avoid having to reengineer the effort at the end of its life cycle. Early systems engineering makes it possible to proactively assess feasibility of meeting user needs, adaptability of new technology, and integration of solutions into regular operations.

STAGES OF INNOVATION PROJECT

The overall stages of innovation project management can be outlined as summarized in the following paragraphs.

Problem Identification

Problem identification is the stage where a need for a proposed project is identified, defined, and justified. A project may be concerned with the development of new products, implementation of new processes, or improvement of existing facilities.

Project Definition

Project definition is the phase at which the purpose of the project is clarified. A *mission statement* is the major output of this stage. For example, a prevailing low level of productivity may indicate a need for a new manufacturing technology. In general, the definition should specify how project management may be used to avoid missed deadlines, poor scheduling, inadequate resource allocation, lack of coordination, poor quality, and conflicting priorities.

Project Planning

A plan represents the outline of the series of actions needed to accomplish a goal. Project planning determines how to initiate a project and execute its objectives. It may be a simple statement of a project goal, or it may be a detailed account of procedures to be followed during the project. Project planning is discussed in detail in Chapter 2. Planning can be summarized as

Objectives
Project definition
Team organization
Performance criteria (time, cost, quality).

Project Organizing

Project organization specifies how to integrate the functions of the personnel involved in a project. Organizing is usually done concurrently with project planning. Directing is an important aspect of project organization. Directing involves guiding and supervising the project personnel. It is a crucial aspect of the management function. Directing requires skillful managers who can interact with subordinates effectively through good communication and motivation techniques. A good project manager will facilitate project success by directing his or her staff, through proper task assignments, toward the project goal.

Workers perform better when there are clearly defined expectations. They need to know how their job functions contribute to the overall goals of the project. Workers should be given some flexibility for self-direction in performing their functions. Individual worker needs and limitations should be recognized by the manager when directing project functions. Directing a project requires skills dealing with motivating, supervising, and delegating.

Resource Allocation

Project goals and objectives are accomplished by allocating resources to functional requirements. Resources can consist of money, people, equipment, tools, facilities, information, skills, and so on. These are usually in short supply. The people needed for a particular task may be committed to other ongoing projects. A crucial piece of equipment may be under the control of another team. Chapter 5 addresses resource allocation in detail.

Project Scheduling

Timeliness is the essence of project management. Scheduling is often the major focus in project management. The main purpose of scheduling is to allocate resources so that the overall project objectives are achieved within a reasonable time span. Project objectives are generally conflicting in nature. For example, minimization of the project completion time and cost are conflicting objectives. That is, one objective is improved at the expense of worsening the other objective. Therefore, project scheduling is a multiple-objective decision-making problem.

In general, scheduling involves the assignment of time periods to specific tasks within the work schedule. Resource availability, time limitations, urgency level, required performance level, precedence requirements, work priorities, technical constraints, and other factors complicate the scheduling process. Thus, the assignment of a time slot to a task does not necessarily ensure that the task will be performed satisfactorily in accordance with the schedule. Consequently, careful control must be developed and maintained throughout the project scheduling process.

Project Tracking and Reporting

This phase involves checking whether or not project results conform to project plans and performance specifications. Tracking and reporting are prerequisites for project control. A properly organized report of the project status will help identify any deficiencies in the progress of the project and help pinpoint corrective actions.

Project Control

Project control requires that appropriate actions be taken to correct unacceptable deviations from expected performance. Control is actuated through measurement, evaluation, and corrective action. Measurement is the process of measuring the relationship between planned performance and actual performance with respect to project objectives. The variables to be measured, the measurement scales, and the measuring approaches should be clearly specified during the planning stage. Corrective actions may involve rescheduling, reallocation of resources, or expedition of task performance. Project control is discussed in detail in Chapter 6. Control involves

Tracking and reporting
Measurement and evaluation
Corrective action (plan revision, rescheduling, updating).

Project Termination

Termination is the last stage of a project. The phaseout of a project is as important as its initiation. The termination of a project should be implemented expeditiously. A project should not be allowed to drag on after the expected completion time. A terminal activity should be defined for a project during the planning phase. An example of a terminal activity may be the submission of a final report, the power on of new equipment, or the signing of a release order. The conclusion of such an activity should be viewed as the completion of the project. Arrangements may be made for follow-up activities that may improve or extend the outcome of the project. These follow-up or spin-off projects should be managed as new projects but with proper input–output relationships within the sequence of projects as serialized below:

Planning → Organizing → Scheduling → Control → Termination

An outline of the functions to be carried out during a project should be made during the planning stage of the project. A model for such an outline is presented hereafter. It may be necessary to rearrange the contents of the outline to fit the specific needs of a project.

Planning

1. Specify project background
   a. Define current situation and process
      i. Understand the process
      ii. Identify important variables
      iii. Quantify variables
   b. Identify areas for improvement
      i. List and discuss the areas
      ii. Study potential strategy for solution
2. Define unique terminologies relevant to the project
   a. Industry-specific terminologies
   b. Company-specific terminologies
   c. Project-specific terminologies
3. Define project goals and objectives
   a. Write mission statement
   b. Solicit inputs and ideas from personnel
4. Establish performance standards
   a. Schedule
   b. Performance
   c. Cost
5. Conduct formal project feasibility study
   a. Determine impact on cost
   b. Determine impact on organization
   c. Determine project deliverables
6. Secure management support.

Organizing

1. Identify project management team
   a. Specify project organization structure
      i. Matrix structure

      ii. Formal and informal structures

      iii. Justify structure

  b. Specify departments involved and key personnel

      i. Purchasing

      ii. Materials management

      iii. Engineering, design, manufacturing, and so on

  c. Define project management responsibilities

      i. Select project manager

      ii. Write project charter

      iii. Establish project policies and procedures

2. Implement Triple C model

  a. Communication

      i. Determine communication interfaces

      ii. Develop communication matrix

  b. Cooperation

      i. Outline cooperation requirements, policies, and procedures

  c. Coordination

      i. Develop work breakdown structure

      ii. Assign task responsibilities

      iii. Develop responsibility chart.

Scheduling (resource allocation)

1. Develop master schedule

  a. Estimate task duration

  b. Identify task precedence requirements

      i. Technical precedence

      ii. Resource-imposed precedence

      iii. Procedural precedence

  c. Use analytical models

      i. Critical path method (CPM)

      ii. Program Evaluation and Review Technique (PERT)

      iii. Gantt chart

      iv. Optimization models.

Control (tracking, reporting, and correction)

1. Establish guidelines for tracking, reporting, and control

  a. Define data requirements

      i. Data categories

      ii. Data characterization

      iii. Measurement scales

    b. Develop data documentation
       i. Data update requirements
      ii. Data quality control
     iii. Establish data security measures
2. Categorize control points
    a. Schedule audit
       i. Activity network and Gantt charts
      ii. Milestones
     iii. Delivery schedule
    b. Performance audit
       i. Employee performance
      ii. Product quality
    c. Cost audit
       i. Cost containment measures
      ii. Percent completion versus budget depletion
3. Identify implementation process
    a. Comparison with targeted schedules
    b. Corrective course of action
       i. Rescheduling
      ii. Reallocation of resources.

Termination (close, phaseout)

1. Conduct performance review
2. Develop strategy for follow-up projects
3. Arrange for personnel retention, release, and reassignment.

Documentation

1. Document project outcome
2. Submit final report
3. Archive report for future reference.

LEAN THINKING AND INNOVATION

Facing a lean period in project management creates value in terms of figuring out how to eliminate or reduce operational waste that is inherent in many human-governed processes. Even in an environment of plenty, thinking lean will make innovation more sustainable. Eliminate waste so that resources can be more effectively assigned to critical areas of need. It is a natural fact that having to make do with limited resources creates opportunities for resourcefulness and innovation, which requires an integrated systems view of what is available and what can be leveraged. The lean principles that are now being embraced by business, industry, and government have been around for a long time. It is just that we are now being forced to implement lean practices due to the escalating shortage of resources. It is unrealistic to expect that problems that have enrooted themselves in different parts of an organization can be solved by a single-point attack. Rather, a systematic probing of all the nooks and corners of the problem must be assessed and tackled in an integrated manner. Like the biblical Joseph, whose life was on the line while interpreting dreams for the Egyptian pharaoh, decision makers cannot afford to misinterpret systems warning signs when managing large and complex projects.

Contrary to the contention in some technocratic circles that budget cuts will stifle innovation, it is a fact that a reduction of resources often forces more creativity in identifying wastes and leveraging opportunities that lie fallow in nooks and crannies of an organization. This is not an issue of wanting more for less. Rather, it is an issue of doing more with less. It is through a systems viewpoint that new opportunities to innovate can be spotted. Necessity does, indeed, spur invention.

INNOVATION DECISION ANALYSIS

Systems decision analysis facilitates a proper consideration of the essential elements of decisions in a project systems environment. These essential elements include the problem statement, information, performance measure, decision model, and an implementation of the decision. The recommended steps are enumerated as follows.

Step 1: Problem Statement

A problem involves choosing between competing, and probably conflicting, alternatives. The components of problem solving in project management include

Describing the problem (goals, performance measures)
Defining a model to represent the problem
Solving the model
Testing the solution
Implementing and maintaining the solution.

Problem definition is very crucial. In many cases, *symptoms* of a problem are more readily recognized than its *cause* and *location*. Even after the problem is accurately identified and defined, a benefit/cost analysis may be needed to determine if the cost of solving the problem is justified.

Step 2: Data and Information Requirements

Information is the driving force for the project decision process. Information clarifies the relative states of past, present, and future events. The collection, storage, retrieval, organization, and processing of raw date are important components for generating information. Without data, there can be no information. Without good information, there cannot be a valid decision. The essential requirements for generating information are

> Ensuring that an effective data collection procedure is followed
> Determining the type and the appropriate amount of data to collect
> Evaluating the data collected with respect to information potential
> Evaluating the cost of collecting the required data.

For example, suppose a manager is presented with a recorded fact that says, "Sales for the last quarter are 10,000 units." This constitutes ordinary data. There are many ways of using the aforementioned data to make a decision, depending on the manager's value system. An analyst, however, can ensure the proper use of the data by transforming it into information, such as "Sales of 10,000 units for the last quarter are within x percent of the targeted value." This type of information is more useful to the manager for decision-making.

Step 3: Performance Measure

A performance measure for the competing alternatives should be specified. The decision maker assigns a perceived worth or value to the available alternatives. Setting measures of performance is crucial to the process of defining and selecting alternatives. Some performance measures commonly used in project management are project cost, completion time, resource usage, and stability in the workforce.

Step 4: Decision Model

A decision model provides the basis for the analysis and synthesis of information and is the mechanism by which competing alternatives are compared. To be effective, a decision model must be based on a systematic and logical framework for guiding project decisions. A decision model can be a verbal, graphical, or mathematical representation of the ideas in the decision-making process. A project decision model should have the following characteristics:

Simplified representation of the actual situation
Explanation and prediction of the actual situation
Validity and appropriateness
Applicability to similar problems.

The formulation of a decision model involves three essential components:

Abstraction: Determining the relevant factors
Construction: Combining the factors into a logical model
Validation: Assuring that the model adequately represents the problem.

The basic types of decision models for project management are described next:

Descriptive models: These models are directed at describing a decision scenario and identifying the associated problem. For example, a project analyst might use a CPM network model to identify bottleneck tasks in a project.

Prescriptive models: These models furnish procedural guidelines for implementing actions. The Triple C approach (Badiru, 2008), for example, is a model that prescribes the procedures for achieving communication, cooperation, and coordination in a project environment.

Predictive models: These models are used to predict future events in a problem environment. They are typically based on historical data about the problem situation. For example, a regression model based on past data may be used to predict future productivity gains associated with expected levels of resource allocation. Simulation models can be used when uncertainties exist in the task durations or resource requirements.

Satisficing models: These are models that provide trade-off strategies for achieving a satisfactory solution to a problem, within given constraints. Goal programming and other multicriteria techniques provide good satisficing solutions. For example, these models are helpful in cases where time limitations, resource shortages, and performance requirements constrain the implementation of a project.

Optimization models: These models are designed to find the best available solution to a problem subject to a certain set of constraints. For example, a linear programming model can be used to determine the optimal product mix in a production environment.

In many situations, two or more of the aforementioned models may be involved in the solution of a problem. For example, a descriptive model might provide insights into the nature of the problem, an optimization model might provide the optimal set of actions to take in solving the problem, a satisficing model might temper the optimal solution with reality, a prescriptive model might suggest the procedures for implementing the selected solution, and a predictive model might forecast a future outcome of the problem scenario.

Step 5: Making the Decision

Using the available data, information, and the decision model, the decision maker will determine the real-world actions that are needed to solve the stated problem. A sensitivity analysis may be useful for determining what changes in parameter values might cause a change in the decision.

Step 6: Implementing the Decision

A decision represents the selection of an alternative that satisfies the objective stated in the problem statement. A good decision is useless until it is implemented. An important aspect of a decision is to specify how it is to be implemented. Selling the decision and the project to management requires a well-organized persuasive presentation. The way a decision is presented can directly influence whether or not it is adopted. The presentation of a decision should include at least the following: an executive summary, technical aspects of the decision, managerial aspects of the decision, resources required to implement the decision, cost of the decision, the time frame for implementing the decision, and the risks associated with the decision.

GROUP DECISION-MAKING FOR INNOVATION

Innovation requires group participation and operation, thus making group decision-making essential. Systems decisions are often complex, diffuse, distributed, and poorly understood. No one person has all the information to make all decisions accurately. As a result, crucial decisions are made by a group of people. Some organizations use outside consultants with appropriate expertise to make recommendations for important decisions. Other organizations set up their own internal consulting groups without having to go outside the organization. Decisions can be made through linear responsibility, in which case one person makes the final decision based on inputs from other people. Decisions can also be made through shared responsibility, in which case, a group of people share the responsibility for making joint decisions. The major advantages of group decision-making are listed as follows:

1. Facilitation of a systems view of the problem environment.
2. Ability to share experience, knowledge, and resources. Many heads are better than one. A group will possess greater collective ability to solve a given decision problem.
3. Increased credibility. Decisions made by a group of people often carry more weight in an organization.
4. Improved morale. Personnel morale can be positively influenced because many people have the opportunity to participate in the decision-making process.
5. Better rationalization. The opportunity to observe other people's views can lead to an improvement in an individual's reasoning process.
6. Ability to accumulate more knowledge and facts from diverse sources.
7. Access to broader perspectives spanning different problem scenarios.
8. Ability to generate and consider alternatives from different perspectives.
9. Possibility for a broad-based involvement, leading to a higher likelihood of support.
10. Possibility for group leverage for networking, communication, and political clout.

In spite of the much-desired advantages, group decision-making does possess the risk of flaws. Some possible disadvantages of group decision-making are listed as follows:

1. Difficulty in arriving at a decision
2. Slow operating time frame
3. Possibility for individuals' conflicting views and objectives
4. Reluctance of some individuals in implementing the decision
5. Potential for power struggle and conflicts among the group
6. Loss of productive employee time
7. Too much compromise may lead to less than optimal group output
8. Risk of one individual dominating the group

9. Overreliance on group process may impede agility of management to make decision fast

10. Risk of dragging feet due to repeated and iterative group meetings.

Brainstorming

Brainstorming is a way of generating many new ideas. In brainstorming, the decision group comes together to discuss alternate ways of solving a problem. The members of the brainstorming group may be from different departments, may have different backgrounds and training, and may not even know one another. The diversity of the participants helps create a stimulating environment for generating different ideas from different viewpoints. The technique encourages free outward expression of new ideas no matter how farfetched the ideas might appear. No criticism of any new idea is permitted during the brainstorming session. A major concern in brainstorming is that extroverts may take control of the discussions. For this reason, an experienced and respected individual should manage the brainstorming discussions. The group leader establishes the procedure for proposing ideas, keeps the discussions in line with the group's mission, discourages disruptive statements, and encourages the participation of all members.

After the group runs out of ideas, open discussions are held to weed out the unsuitable ones. It is expected that even the rejected ideas may stimulate the generation of other ideas, which may eventually lead to other favored ideas. Guidelines for improving brainstorming sessions are presented as follows:

Focus on a specific decision problem.
Keep ideas relevant to the intended decision.
Be receptive to all new ideas.
Evaluate the ideas on a relative basis after exhausting new ideas.
Maintain an atmosphere conducive to cooperative discussions.
Maintain a record of the ideas generated.

Delphi Method

The traditional approach to group decision-making is to obtain the opinion of experienced participants through open discussions. An attempt is made to reach a consensus among the participants. However, open group discussions are often biased because of the influence of subtle intimidation from dominant individuals. Even when the threat of a dominant individual is not present, opinions may still be swayed by group pressure. This is called the "bandwagon effect" of group decision-making.

The Delphi method attempts to overcome these difficulties by requiring individuals to present their opinions anonymously through an intermediary. The method differs from the other interactive group methods because it eliminates face-to-face confrontations. It was originally developed for forecasting applications, but it has been modified in various ways for application to different types of decision-making. The method can be quite useful for project management decisions. It is particularly effective when decisions must be based on a broad set of factors. The Delphi method is normally implemented as follows:

1. *Problem definition:* A decision problem that is considered significant is identified and clearly described.

2. *Group selection:* An appropriate group of experts or experienced individuals is formed to address the particular decision problem. Both internal and external experts may be involved in the Delphi process. A leading individual is appointed to serve as the administrator of the decision process. The group may operate through the mail or gather together in a room. In either case, all opinions are expressed anonymously on paper. If the group meets in the same room, care should be taken to provide enough room so that each member does not have the feeling that someone may accidentally or deliberately observe their responses.

3. *Initial opinion poll:* The technique is initiated by describing the problem to be addressed in unambiguous terms. The group members are requested to submit a list of major areas of concern in their specialty areas as they relate to the decision problem.

4. *Questionnaire design and distribution:* Questionnaires are prepared to address the areas of concern related to the decision problem. The written responses to the questionnaires are collected and organized by the administrator. The administrator aggregates the responses in a statistical format. For example, the average, mode, and median of the responses may be computed. This analysis is distributed to the decision group. Each member can then see how his or her responses compare with the anonymous views of the other members.

5. *Iterative balloting:* Additional questionnaires based on the previous responses are passed to the members. The members submit their responses again. They may choose to alter or not to alter their previous responses.

6. *Silent discussions and consensus:* The iterative balloting may involve anonymous written discussions of why some responses are correct or incorrect. The process is continued until a consensus is

reached. A consensus may be declared after five or six iterations of the balloting or when a specified percentage (e.g., 80%) of the group agrees on the questionnaires. If a consensus cannot be declared on a particular point, it may be displayed to the whole group with a note that it does not represent a consensus.

In addition to its use in technological forecasting, the Delphi method has been widely used in other general decision-making. Its major characteristics of anonymity of responses, statistical summary of responses, and controlled procedure make it a reliable mechanism for obtaining numeric data from subjective opinion. The major limitations of the Delphi method are as follows:

1. Its effectiveness may be limited in cultures where strict hierarchy, seniority, and age influence decision-making processes.
2. Some experts may not readily accept the contribution of nonexperts to the group decision-making process.
3. Since opinions are expressed anonymously, some members may take the liberty of making ludicrous statements. However, if the group composition is carefully reviewed, this problem may be avoided.

Nominal Group Technique

The nominal group technique is a silent version of brainstorming. It is a method of reaching consensus. Rather than asking people to state their ideas aloud, the team leader asks each member to jot down a minimum number of ideas, for example, five or six. A single list of ideas is then written on a chalkboard for the whole group to see. The group then discusses the ideas and weeds out some iteratively until a final decision is made. The nominal group technique is easier to control. Unlike brainstorming where members may get into shouting matches, the nominal group technique permits members to silently present their views. In addition, it allows introversive members to contribute to the decision without the pressure of having to speak out too often.

In all of the group decision-making techniques, an important aspect that can enhance and expedite the decision-making process is to require that members review all pertinent data before coming to the group meeting. This will ensure that the decision process is not impeded by trivial preliminary discussions. Some disadvantages of group decision-making are as follows:

1. Peer pressure in a group situation may influence a member's opinion or discussions.
2. In a large group, some members may not get to participate effectively in the discussions.

3. A member's relative reputation in the group may influence how well his or her opinion is rated.
4. A member with a dominant personality may overwhelm the other members in the discussions.
5. The limited time available to the group may create a time pressure that forces some members to present their opinions without fully evaluating the ramifications of the available data.
6. It is often difficult to get all members of a decision group together at the same time.

Despite the noted disadvantages, group decision-making definitely has many advantages that may nullify the shortcomings. The advantages as presented earlier will have varying levels of effect from one organization to another. The Triple C principle (Badiru, 2008), may be used to improve the success of decision teams. Team work can be enhanced in group decision-making by adhering to the following guidelines:

1. Get a willing group of people together.
2. Set an achievable goal for the group.
3. Determine the limitations of the group.
4. Develop a set of guiding rules for the group.
5. Create an atmosphere conducive to group synergism.
6. Identify the questions to be addressed in advance.
7. Plan to address only one topic per meeting.

For major decisions and long-term group activities, arrange for team training that allows the group to learn the decision rules and responsibilities together. The steps for the nominal group technique are as follows:

1. Silently generate ideas, in writing.
2. Record ideas without discussion.
3. Conduct group discussion for clarification of meaning, not argument.
4. Vote to establish the priority or rank of each item.
5. Discuss vote.
6. Cast final vote.

Interviews, Surveys, and Questionnaires

Interviews, surveys, and questionnaires are important information gathering techniques. They also foster cooperative working relationships. They encourage direct participation and inputs into project decision-making processes.

They provide an opportunity for employees at the lower levels of an organization to contribute ideas and inputs for decision-making. The greater the number of people involved in the interviews, surveys, and questionnaires, the more valid the final decision. The following guidelines are useful for conducting interviews, surveys, and questionnaires to collect data and information for project decisions:

1. Collect and organize background information and supporting documents on the items to be covered by the interview, survey, or questionnaire.
2. Outline the items to be covered and list the major questions to be asked.
3. Use a suitable medium of interaction and communication: telephone, fax, electronic mail, face to face, observation, meeting venue, poster, or memo.
4. Tell the respondent the purpose of the interview, survey, or questionnaire, and indicate how long it will take.
5. Use open-ended questions that stimulate ideas from the respondents.
6. Minimize the use of yes or no type of questions.
7. Encourage expressive statements that indicate the respondent's views.
8. Use the who, what, where, when, why, and how approach to elicit specific information.
9. Thank the respondents for their participation.
10. Let the respondents know the outcome of the exercise.

Multivote

Multivoting is a series of votes used to arrive at a group decision. It can be used to assign priorities to a list of items. It can be used at team meetings after a brainstorming session has generated a long list of items. Multivoting helps reduce such long lists to a few items, usually three to five. The steps for multivoting are as follows:

1. Take a first vote. Each person votes as many times as desired but only once per item.
2. Circle the items receiving a relatively higher number of votes (i.e., majority vote) than the other items.
3. Take a second vote. Each person votes for a number of items equal to one-half the total number of items circled in Step 2. Only one vote per item is permitted.

4. Repeat Steps 2 and 3 until the list is reduced to three to five items depending on the needs of the group. It is not recommended to multivote down to only one item.
5. Perform further analysis of the items selected in Step 4, if needed.

SUMMARY

As can be seen in the preceding discussions, innovation can be a challenge to manage. For that reason, classical systems approach and formal project management processes are essential for instituting and sustaining innovation. Innovation doesn't just happen, it must be managed as a rigorous project. The absence of a structured project management approach could heighten the risk for innovation failure.

REFERENCES

Badiru, A.B. (2008). *Triple C Model of Project Management: Communication, Cooperation, and Coordination*, Boca Raton, FL: CRC Press/Taylor & Francis Group.

Badiru, A.B. (2019). *Project Management: Systems, Principles, and Applications*, 2nd ed. Boca Raton, FL: CRC Press/Taylor & Francis Group.

Troxler, J.W. and Blank, L. (1989). A comprehensive methodology for manufacturing system evaluation and comparison, *Journal of Manufacturing Systems*, 8(3), 176–183.

DEJI Systems Model for Innovation

3

INTRODUCTION

"A new way of doing things" is one of the definitions of innovation presented earlier. Whether it is this definition or any other definition, innovation must have a buy-in from stakeholders and it must be integrated into the operating environment. The DEJI systems model (Badiru, 2012, 2019) is a good tool for ensuring that the proposed innovation fits the operating environment of the organization. The DEJI systems model is applicable for innovation **Design, Evaluation, Justification**, and **Integration**. Figure 3.1 illustrates the DEJI model.

Several factors related to innovation are amenable to the application of the DEJI model. Some of these are discussed in the sections that follow.

Innovation Quality Management

Quality is a measure of customer satisfaction and a product's "fit-for-use" status. To perform its intended functions, a product must provide a balanced level of satisfaction to both the producer and the customer. For that purpose, this author presents the following comprehensive definition of quality:

> Quality refers to an equilibrium level of functionality possessed by a product or service based on the producer's capability and the customer's needs.

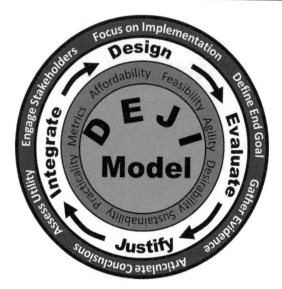

FIGURE 3.1   Elements of the DEJI systems model for innovation.

Based on the above definition, quality refers to the combination of character-istics of a product, process, or service that determines the product's ability to satisfy specific needs. Quality is a product's ability to conform to specifi-cations, where specifications represent the customer's needs or government regulations. The attainment of quality in a product is the responsibility of every employee in an organization, and the production and preservation of quality should be a commitment that extends all the way from the producer to the customer. Products that are designed to have high quality cannot main-tain the inherent quality at the user's end of the spectrum if they are not used properly.

The functional usage of a product should match the functional specifica-tions for the product within the prevailing usage environment. The ultimate judge for the quality of a product, however, is the perception of the user, and differing circumstances may alter that perception. A product that is perceived as being of high quality for one purpose at a given time may not be seen as hav-ing acceptable quality for another purpose in another time frame. Industrial quality standards provide a common basis for global commerce. Customer satisfaction or production efficiency cannot be achieved without product standards. Regulatory, consensus, and contractual requirements should be taken into account when developing product standards driven by innovation. These are described below.

Regulatory Standards

This refers to standards that are imposed by a governing body, such as a government agency. All firms within the jurisdiction of the agency are required to comply with the prevailing regulatory standards.

Consensus Standards

This refers to a general and mutual agreement between companies to abide by a set of self-imposed standards.

Contractual Standards

Contractual standards are imposed by the customer based on case-by-case or order-by-order needs. Most international standards will fall into the category of consensus standards, simply because a lack of an international agreement often leads to trade barriers.

Innovative Product Design

The initial step in any manufacturing effort is the development of a manufacturable and marketable product. An analysis of what is required for a design and what is available for the design should be conducted in the planning phase of a design project. The development process must cover analyses of the product configuration, the raw materials required, production costs, and potential profits. Design engineers must select appropriate materials, the product must be expected to operate efficiently for a reasonable length of time (reliability and durability), and it must be possible to manufacture the product at a competitive cost. The design process will be influenced by the required labor skills, production technology, and raw materials. Product planning is substantially influenced by the level of customer sophistication, enhanced technology, and competition pressures. These are all project-related issues that can be enhanced by project management. The designer must recognize changes in all these factors and incorporate them into the design process. Design project management provides a guideline for the initiation, implementation, and termination of a design effort. It sets guidelines for specific design objectives, structure, tasks, milestones, personnel, cost, equipment, performance, and problem resolutions. The steps involved include planning, organizing, scheduling, and control. The availability of technical expertise within an organization and outside of it should be reviewed. The primary question of whether or not a design is needed at all should be addressed. The "make" or "buy," "lease" or "rent," and "do nothing" alternatives to a proposed design should be among the considerations.

In the initial stage of design planning, the internal and external factors that may influence the design should be determined and given relative weights according to priority. Examples of such influential factors include organizational goals, labor situations, market profile, expected return on design investment, technical manpower availability, time constraints, state of the technology, and design liabilities. The desired components of a design plan include summary of the design plan, design objectives, design approach, implementation requirements, design schedule, required resources, available resources, design performance measures, and contingency plans.

Design Feasibility

The feasibility of a proposed design can be ascertained in terms of technical factors, economic factors, or both. A feasibility study is documented with a report showing all the ramifications of the design. A report of the design's feasibility should cover statements about the need, the design process, the cost feasibility, and the design effectiveness. The need for a design may originate from within the organization, from another organization, from the public, or from the customer. Pertinent questions for design feasibility review include: Is the need significant enough to warrant the proposed design? Will the need still exist by the time the design is finished? What are the alternate means of satisfying the need? What technical interfaces are required for the design? What is the economic impact of the need? What is the return, financially, on the design change?

A Design Breakdown Structure (DBS) is a flowchart of design tasks required to accomplish design objectives. Tasks that are contained in the DBS collectively describe the overall design. The tasks may involve hardware products, software products, services, and information. The DBS helps to describe the link between the end objective and its components. It shows design elements in the conceptual framework for the purposes of planning and control. The objective of developing a DBS is to study the elemental components of a design project in detail, thus permitting a "divide and conquer" approach. Overall design planning and control can be significantly improved by using DBS. A large design may be decomposed into smaller sub-designs, which may, in turn, be decomposed into task groups. Definable sub-goals of a design problem may be used to determine appropriate points at which to decompose the design.

Individual components in a DBS are referred to as *DBS elements* and the hierarchy of each is designated by a level identifier. Elements at the same level of subdivision are said to be of the same DBS level. Descending levels provide increasingly detailed definition of design tasks. The complexity of a design and the degree of control desired are used to determine the number of levels to have in a DBS. Level I of a DBS contains only the final design purpose. This item should be identifiable directly as an organizational goal. Level II contains the major subsections of the design. These subsections are usually identified by their contiguous location or by their related purpose. Level III contains definable components of the Level II subsections. Subsequent levels are constructed in more specific details depending on the level of control desired. If a complete DBS becomes too crowded, separate DBSs may be drawn for the Level II components, for example. A specification of design (SOD) should accompany the DBS. A statement of design is a narrative of the design to be generated. It should include the objectives of the design, its nature, the resource requirements, and a tentative schedule. Each DBS element is assigned a code (usually numeric) that is used for the element's identification throughout the design life cycle.

Design Stages

The guidelines for the various stages in the life cycle of a design can be summarized in the following way:

1. *Definition of design problem:* Define problem and specify the importance of the problem, emphasize the need for a focused design problem, identify designers willing to contribute expertise to the design process, and disseminate the design plan.

2. *Personnel assignment:* The design group and the respective tasks should be announced and a design manager should be appointed to oversee the design effort.

3. *Design initiation:* Arrange organizational meeting, discuss general approach to the design problem, announce specific design plan, and arrange for the use of required hardware and tools.

4. *Design prototype:* Develop a prototype design, test an initial implementation, and learn more about the design problem from test results.

5. *Full design development:* Expand the prototype design and incorporate user requirements.

6. *Design verification:* Get designers and potential users involved, ensure that the design performs as designed, and modify the design as needed.

7. *Design validation:* Ensure that the design yields the expected outputs. Validation can address design performance level, deviation from expected outputs, and the effectiveness of the solution to the problem.

8. *Design integration:* Implement the full design, ensure the design is compatible with existing designs and manufacturing processes, and arrange for design transfer to other processes.

9. *Design feedback analysis:* What are the key lessons from the design effort? Were enough resources assigned? Was the design completed on time? Why or why not?

10. *Design maintenance:* Arrange for continuing technical support of the design and update design as new information or technology becomes available.

11. *Design documentation:* Prepare full documentation of the design and document the administrative process used in generating the design.

Cultural and Social Compatibility Issues

Cultural infeasibility is one of the major impediments to outsourcing innovation in a wide-open market. The business climate can be very volatile. This volatility, coupled with cultural limitations, creates problematic operational, particularly in an emerging technology. The pervasiveness of online transactions overwhelms the strict cultural norms in many markets. The cultural feasibility of information-based outsourcing needs to be evaluated from the standpoint of where information originates, where it is intended to go, and who comes into contact with the information. For example, the revelation of personal information is frowned upon in many developing countries, where there may be an interest in outsourced innovation engagements. Consequently, this impedes the collection, storage, and distribution of workforce information that may be vital to the success of outsourcing. For outsourcing to be successfully implemented in such settings, assurances must be incorporated into the hardware and software implementations so as to conciliate the workforce. Accidental or deliberate mismanagement of information is a more worrisome aspect of IT than it is in the Western world, where enhanced techniques are available to correct information errors. What is socially acceptable in the outsourcing culture may not be acceptable in the receiving culture and vice versa.

Administrative Compatibility

Administrative or managerial feasibility involves the ability to create and sustain an infrastructure to support an operational goal. Should such an infrastructure not be in existence or unstable, then we have a case of administrative infeasibility. In developing countries, a lack of trained manpower precludes a stable infrastructure for some types of industrial outsourcing. Even where trained individuals are available, the lack of coordination makes it almost impossible to achieve a collective and dependable workforce. Systems that are designed abroad for implementation in a different setting frequently get bogged down when imported into a developing environment that is not conducive for such systems. Differences in the perception of ethics are also an issue of concern in an outsource location. A lack of administrative vision and limited managerial capabilities limit the ability of outsource managers in developing countries. Both the physical and conceptual limitations on technical staff lead to administrative infeasibility that must be reckoned with. Overzealous entrepreneurs are apt to jump on opportunities to outsource production without a proper assessment of the capabilities of the receiving organization. Most often than not, outsourcing organizations don't fully understand the local limitations. Some organizations take the risk of learning as they go, without adequate prior preparation.

Technical Compatibility

Hardware maintenance and software upgrade are, perhaps, the two most noticeable aspects of technical infeasibility of information technology in a developing country. The mistake is often made that once you install IT and all its initial components, you have the system for life. This is very far from the truth. The lack of proximity to the source of hardware and software enhancement makes this situation particularly distressing in a developing country. The technical capability of the personnel as well as the technical status of the hardware must be assessed in view of the local needs. Doing an overkill on the infusion of IT just for the sake of keeping up is as detrimental as doing nothing at all.

Workforce Integration Strategies

Any outsourcing enterprise requires adapting from one form of culture to another. The implementation of a new technology to replace an existing (or a nonexistent) technology can be approached through one of several cultural adaptation options. Below are some suggestions:

Parallel interface: The host culture and the guest culture operate concurrently (side by side), with mutual respect on either side.

Adaptation interface: This is the case where either the host culture or the guest culture makes conscious effort to adapt to each other's ways. The adaptation often leads to new (but not necessarily enhanced) ways of thinking and acting.

Superimposition interface: The host culture is replaced (annihilated or relegated) by the guest culture. This implies cultural imposition on local practices and customs. Cultural incompatibility, for the purpose of business goals, is one reason to adopt this type of interface.

Phased interface: Modules of the guest culture are gradually introduced to the host culture over a period of time.

Segregated interface: The host and guest cultures are separated both conceptually and geographically. This used to work well in colonial days. But it has become more difficult with modern flexibility of movement and communication facilities.

Pilot interface: The guest culture is fully implemented on a pilot basis in a selected cultural setting in the host country. If the pilot implementation works with good results, it is then used to leverage further introduction to other localities.

Hybridization of Innovation Cultures

The increased interface of cultures through industrial outsourcing is gradually leading to the emergence of hybrid cultures in many developing countries. A hybrid culture derives its influences from diverse factors, where there are differences in how the local population views education, professional loyalty, social alliances, leisure pursuits, and information management. A hybrid culture is, consequently, not fully embraced by either side of the cultural divide. This creates a big challenge to managing outsourcing projects. Figure 3.2 illustrates innovation interfaces that may influence operating cultures in business, industry, academia, defense, or government.

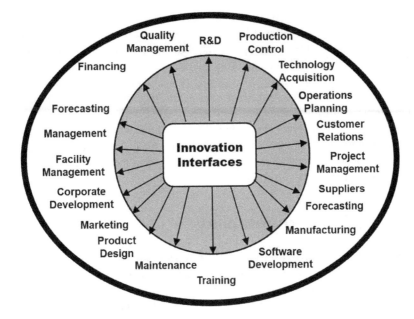

FIGURE 3.2   Organizational interfaces of innovation.

INNOVATION QUALITY INTERFACES

Quality is at the intersection of efficiency, effectiveness, and productivity. Efficiency provides the framework for quality in terms of resources and inputs required to achieve the desired level of quality. Effectiveness comes into play with respect to the application of product quality to meet specific needs and requirements of an organization. Productivity is an essential factor in the pursuit of quality as it relates to the throughput of a production system. To achieve the desired levels of quality, efficiency, effectiveness, and productivity, a new research framework must be adopted. In this column, we present a potential quality enhancement model for quality DEJI based on the product development application presented by Badiru (2012). The model is relevant for research efforts in quality engineering and technology applications.

This second installment of the research column on quality insights continues the contribution set out in the inaugural column in the September 2014 issue. Several aspects of quality must undergo rigorous research along the realms of both quantitative and qualitative characteristics. Many times, quality is taken for granted and the flaws only come out during the implementation stage, which may be too late to rectify. The growing trend in product recalls is a symptom of a priori analysis of the sources and implications of quality at the product conception stage. This column advocates the use of the DEJI model for enhancing quality design, quality evaluation, quality justification, and quality integration through hierarchical and stage-by-stage processes.

Better quality is achievable, and there is always room for improvement in the quality of products and services. But we must commit more efforts to the research at the outset of the product development cycle. Even the human elements of the perception of quality can benefit from more directed research from a social and behavioral sciences point of view.

Innovation Accountability

Throughout history, engineering has answered the call of the society to address specific challenges. With such answers comes a greater expectation of professional accountability. Consider the level of social responsibility that existed during the time of the Code of Hammurabi. Two of the laws are echoed below:

Hammurabi's Law 229:
>   If a builder build a house for someone, and does not construct it properly, and the house which he built fall in and kill its owner, then that builder shall be put to death.

Hammurabi's Law 230:
>   If it kills the son of the owner the son of that builder shall be put to death.

These are drastic measures designed to curb professional dereliction of duty and enforce social responsibility with particular focus on product quality. Research and education must play bigger and more direct roles in the design, practice, and management of quality and present modern aspects of social responsibility in the context of day-to-day personal and professional activities. The global responsibility of the greater society is essential with respect to world development challenges covering the global economy, human development, global governance, and social relationships. Quality is the common theme in the development challenges. Focusing on the emerging field of Big Data, we should advocate engineering education collaboration, which aligns well with data-intensive product development. With the above principles as possible tenets for better research, education, and practice of quality in engineering and technology, this chapter suggests the DEJI model as a potential methodology.

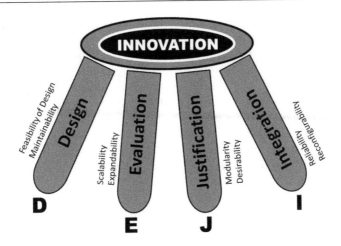

FIGURE 3.3   Four-legged stool of DEJI systems model for innovation.

The DEJI Model

The DEJI model encourages the practice of building quality into a product right from the beginning so that the product integration stage can be more successful. Figure 3.3 shows the four-legged model for an overall systems quality with respect to DEJI.

Design of Quality

The design of quality in product development should be structured to follow point-to-point transformations. A good technique to accomplish this is the use of state-space transformation, with which we can track the evolution of a product from the concept stage to a final product stage. For the purpose of product quality design, the following definitions are applicable:

> *Product state:* A state is a set of conditions that describe the product at a specified point in time. The *state* of a product refers to a performance characteristic of the product which relates input to output such that a knowledge of the input function over time and the state of the product at time $t = t_0$ determines the expected output for $t \geq t_0$. This is particularly important for assessing where the product stands in the context of new technological developments and the prevailing operating environment.

Product state space: A product *state space* is the set of all possible states of the product life cycle. State-space representation can solve product design problems by moving from an initial state to another state and eventually to the desired end-goal state. The movement from state to state is achieved by means of actions. A goal is a description of an intended state that has not yet been achieved. The process of solving a product problem involves finding a sequence of actions that represents a solution path from the initial state to the goal state. A state-space model consists of state variables that describe the prevailing condition of the product. The state variables are related to inputs by mathematical relationships. Examples of potential product state variables include schedule, output quality, cost, due date, resource, resource utilization, operational efficiency, productivity throughput, and technology alignment. For a product described by a system of components, the state-space representation can follow the quantitative metric below:

$$\mathbf{Z} = \mathbf{f}(z,x); \mathbf{Y} = \mathbf{g}(z,x)$$

where $\mathbf{f}$ and $\mathbf{g}$ are vector-valued functions. The variable $\mathbf{Y}$ is the output vector, while the variable x denotes the inputs. The state vector $\mathbf{Z}$ is an intermediate vector relating x to y. In generic terms, a product is transformed from one state to another by a driving function that produces a transitional relationship given by

$$S_s = f\left(x \middle| S_p\right) + e,$$

where S_s = subsequent state, x = state variable, S_p = the preceding state, and e = error component.

The function f is composed of a given action (or a set of actions) applied to the product. Each intermediate state may represent a significant milestone in the project. Thus, a descriptive state-space model facilitates an analysis of what actions to apply in order to achieve the next desired product state. A graphical representation can be developed for a product transformation from one state to another through the application of human or machine actions. This simple representation can be expanded to cover several components within the product information framework. Hierarchical linking of product elements provides an expanded transformation structure. The product state can be expanded in accordance with implicit requirements. These requirements might include grouping of design elements, linking precedence requirements (both technical and procedural), adapting to new technology developments, following required communication links, and accomplishing reporting requirements. The actions to be taken at each state depend on the prevailing product conditions. The nature of subsequent alternate states depends on what actions are implemented. Sometimes there are multiple paths that can lead to the desired end result. At other times, there exists only one unique path to the desired objective. In conventional practice, the characteristics of the future states can only be recognized after the fact, thus, making it impossible to develop adaptive plans. In the implementation of the **DEJI** model, adaptive plans can be achieved because the events occurring within and outside the product state boundaries can be taken into account.

If we describe a product by P state variables s_i, then the composite state of the product at any given time can be represented by a vector $\mathbf{S}$ containing P elements. That is,

$$\mathbf{S} = \{s_1, s_2, \ldots, s_P\}$$

The components of the state vector could represent either quantitative or qualitative variables (e.g., cost, energy, color, time). We can visualize every state vector as a point in the state space of the product. The representation is unique since every state vector corresponds to one and only one point in the state space. Suppose we have a set of actions (transformation agents) that we can apply to the product information so as to change it from one state to another within the project state space. The transformation will change a state vector into another state vector. A transformation may be a change in raw material or a change in design approach. The number of transformations available for a product characteristic may be finite or unlimited. We can construct trajectories that describe the potential states of a product evolution as we apply successive transformations with respect to technology forecasts. Each transformation may be repeated as many times as needed. Given an initial state $\mathbf{S}_0$, the sequence of state vectors is represented by the following:

$$\mathbf{S}_n = T_n(\mathbf{S}_{n-1}).$$

The state-by-state transformations are then represented as $\mathbf{S}_1 = T_1(\mathbf{S}_0)$; $\mathbf{S}_2 = T_2(\mathbf{S}_1)$; $\mathbf{S}_3 = T_3(\mathbf{S}_2)$; $\ldots$; $\mathbf{S}_n = T_n(\mathbf{S}_{n-1})$. The final state, $\mathbf{S}_n$, depends on the initial state $\mathbf{S}$ and the effects of the actions applied.

Evaluation of Quality

A product can be evaluated on the basis of cost, quality, schedule, and meeting requirements. There are many quantitative metrics that can be used in evaluating a product at this stage. Learning curve productivity is one relevant technique that can be used because it offers an evaluation basis of a product with respect to the concept of growth and decay. The half-life extension (Badiru, 2012) of the basic learning is directly applicable because the half-life of the technologies going into a product can be considered. In today's technology-based operations, retention of learning may be threatened by fast-paced shifts in operating requirements. Thus, it is of interest to evaluate the half-life properties of new technologies as they impact the overall product quality. Information about the half-life can tell us something about the sustainability of learning-induced technology performance. This is particularly useful for designing products whose life cycles stretch into the future in a high-tech environment.

Justification of Quality

We need to justify an innovation program on the basis of quantitative value assessment. The systems value model (SVM) is a good quantitative technique that can be used here for innovation justification on the basis of value. The model provides a heuristic decision aid for comparing project alternatives. It is presented here again for the present context. Value is represented as a deterministic vector function that indicates the value of tangible and intangible attributes that characterize the project. It is represented as $V = f\left(A_1, A_2, ..., A_p\right)$, where V is the assessed value and the A values are quantitative measures or attributes. Examples of product attributes are quality, throughput, manufacturability, capability, modularity, reliability, interchangeability, efficiency, and cost performance. Attributes are considered to be a combined function of factors. Examples of product factors are market share, flexibility, user acceptance, capacity utilization, safety, and design functionality. Factors are themselves considered to be composed of indicators. Examples of indicators are debt ratio, acquisition volume, product responsiveness, substitutability, lead time, learning curve, and scrap volume. By combining the above definitions, a composite measure of the operational value of a product can be quantitatively assessed. In addition to the quantifiable factors, attributes, and indicators that impinge upon overall project value, the human-based subtle factors should also be included in assessing overall project value.

Earned Value Technique for Innovation

Value is synonymous with quality. Thus, the contemporary earned value technique is relevant for "earned quality" analysis. This is a good analytical technique to use for the justification stage of the DEJI model. This will impact cost, quality, and schedule elements of product development with respect to value creation. The technique involves developing important diagnostic values for each schedule activity, work package, or control element. The variables are: PV: planned value; EV: earned value; AC: actual cost; CV: cost variance; SV: schedule variance; EAC: estimate at completion; BAC: budget at completion; and ETC: estimate to complete. This analogical relationship is a variable research topic for quality engineering and technology applications.

Integration of Quality

Without being integrated, a system will be in isolation and it may be worthless. We must integrate all the elements of a system on the basis of alignment of functional goals. The overlap of systems for integration purposes can conceptually be viewed as projection integrals by considering areas bounded by the common elements of sub-systems. Quantitative metrics can be applied at this stage for effective assessment of the product state. Trade-off analysis is essential in quality integration. Pertinent questions include the following:

What level of trade-offs on the level of quality is tolerable?
What is the incremental cost of higher quality?
What is the marginal value of higher quality?
What is the adverse impact of a decrease in quality?

What is the integration of quality of time? In this respect, an integral of the form below may be suitable for further research:

$$I = \int_{t_1}^{t_2} f(q)\,dq,$$

where I = integrated value of quality, $f(q)$ = functional definition of quality, t_1 = initial time, and t_2 = final time within the planning horizon.

Presented below are guidelines and important questions relevant for quality integration:

- What are the unique characteristics of each component in the integrated system?
- How do the characteristics complement one another?
- What physical interfaces exist among the components?
- What data/information interfaces exist among the components?
- What ideological differences exist among the components?

- What are the data flow requirements for the components?
- What internal and external factors are expected to influence the integrated system?
- What are the relative priorities assigned to each component of the integrated system?
- What are the strengths and weaknesses of the integrated system?
- What resources are needed to keep the integrated system operating satisfactorily?
- Which organizational unit has primary responsibility for the integrated system?

The proposed approach of the DEJI model will facilitate a better alignment of product technology with future development and needs. The stages of the model require research for each new product with respect to DEJI. Existing analytical tools and techniques can be used at each stage of the model.

SUMMARY

Innovation is an integrative process that must be evaluated on a stage-by-stage approach. This requires research, education, and implementation strategies that consider several pertinent factors. This chapter suggests the DEJI model, which has been used successfully for product development applications, as a viable methodology for innovation design, evaluation, justification, and integration. The presentation in this chapter is intended as a source to spark the interest of researchers to apply this tool in new innovation efforts.

REFERENCES

Badiru, A.B. (2012). Application of the DEJI model for aerospace product integration, *Journal of Aviation and Aerospace Perspectives (JAAP)*, 2(2), 20–34.
Badiru, A.B. (2019). *Systems Engineering Models: Theory, Methods, and Applications.* Boca Raton, FL: CRC Press/Taylor & Francis Group.

Umbrella Model for Innovation

4

MODEL-BASED INNOVATION

Innovation is presently the hottest topic in business, industry, academia, and the government. In response to a prevailing priority of business and industry to drive innovation, this chapter introduces a research study on the development of a theory of innovation from the perspective of how people work and collaborate in the pursuit of innovation within the defense acquisition framework. The focus of innovation ranges from the acquisition of technology products, services, operational processes as well as workforce talent. This justifies using a systems framework for the methodology development in this chapter. The *Umbrella Theory of Innovation* (Badiru, 2019) is generally applicable for innovation pursuits in diverse operational environments in business, industry, government, the military, and academia.

Innovation is currently one of the most embraced words in business, industry, academia, government, and the military. However, it is, perhaps, the most misunderstood in terms of operational manifestation. Badiru (2019) introduced a foundational process of developing a theory of innovation from the perspective of how people work and collaborate in the pursuit of innovation within the defense acquisition framework. As a specific focus, this chapter uses a systems theoretic approach to develop techniques and strategies for driving innovation throughout an innovation technology acquisition life cycle. Of particular interest is the view of the acquisition system as a learning system. A learning system is a sustainable system. The national goal of achieving a progressive and sustainable acquisition can be advanced through systems theoretic methodologies. The methodology of the model considers how people communicate, cooperate, coordinate, and collaborate to actualize the concepts and ideas embedded in innovation initiatives. What does it mean to pursue and actualize innovation? The answer is a mix of quantitative and qualitative processes in any organization.

The theme of innovation is presently sweeping through business and industry. Organizations in business, industry, government, academia, and the military are all embracing innovation with different flavors of conceptual and practical pursuits. The process of managing and actualizing innovation can be ambiguous and intractable because innovation is not a tangible product that can be assessed with traditional performance metrics. From a control perspective, innovation is nothing more than using a rigorous management process to link concepts and ideas to some desired output. That output will be in the form one of three possibilities:

- Product (a physical output of innovation)
- Service (a provision resulting from innovation)
- Result (a desired outcome of innovation).

Essentially, innovation is the pathway from an initial conceptual point to a discernible end point as illustrated in Figure 4.1. The ingredients of innovation are the following:

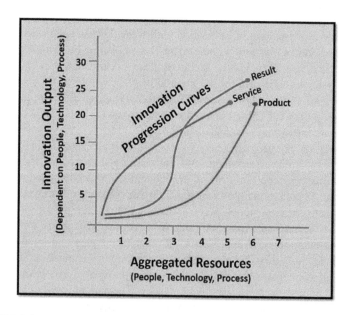

FIGURE 4.1   Innovation progression curves to desired outputs.

1. Technology framework, upon which innovation is expected to happen
2. Workforce, upon whose education, training, and experience, innovation is supposed to rest
3. Operational process, on the basis of which actions take place to make innovation happen.

Successful innovation is predicated on a solid foundation and interplay of people, technology, and process. In the hypothetical rendition of Figure 4.1, the dependent variable is "innovation output." The independent variable is the aggregated resource level, which is composed of people, technology, and process. Those resource elements could, themselves, be dependent on other organizational assets, thereby creating hierarchical embedment of interrelated elements in a systems structure. The end point of each curve in Figure 4.1 represents either a desired result, an expected service, or a required product.

Innovation is multidimensional and has been addressed from different perspectives in the literature. Keeley et al. (2013) discuss ten types of innovation as enumerated below:

1. Profit model
2. Network
3. Structure
4. Process
5. Product performance
6. Product system
7. Service
8. Channel
9. Brand
10. Customer engagement.

Innovation types 1–4 are categorized as falling under the group heading of "Configuration." Types 5 and 6 are grouped under "Offering." Innovation types 7–10 fall under "Experience." Configuration-based innovation is focused on the inherent workings of an enterprise and its business system. Offering-based innovation is focused on the enterprise's core product or service, or a collection of its products and services. Experience-based innovation is focused on more customer-directed elements of an enterprise and its business system. The authors emphasize that the above categorization does not imply process timeline, sequencing, or hierarchy among the types of innovation. In fact, any combination of types can be present in any pursuit of innovation. Thus, the framework embraced by any innovation-centric organization can be anchored and initiated at any of the ten types. This free-flow concept fits the systems theoretic premise of this chapter. Viewed as a system, the pursuit of innovation can have a variety of elements, working together, to produce better overall output for the organization.

Voehl et al. (2019) present a collection of topics that can make up the Innovation Body of Knowledge (IBOK). The concept, tools, and techniques presented in their book reinforce the need to take a systems view of innovation. Coverages in the chapter include preparing the organization for innovation, promoting and communicating innovation, creativity for entrepreneurship both personal and corporate, innovation process model, business readiness for innovation, building organizational foundation for innovation, interdisciplinary approach to TRIZ (Theory of Inventive Problem Solving) and STEM (science, technology, engineering, and mathematics), and intellectual property management for innovation. The diversity of topics in this chapter and other literature sources confirm that innovation is not just one "thing." A systems theoretic approach is, indeed, required to analyze and synthesize all the factors involved in innovation.

A broad, intensive, and detailed review of the literature on innovation confirms the multifaceted nuances and requirements for driving innovation in a defense acquisition system. There are several case examples of where and how innovation is desired in the US Military Strength. Satell (2017) presents what he calls a playbook for navigating a disruptive age for the purpose of following a mapping scheme through the latest technological developments. Hamel (2012) covers what matters in innovation pursuits in terms of values, passion, adaptability, and ideology in an innovation environment. Degraff and Degraff (2017) highlight the essentiality of constructive conflict in the pursuit of innovation. Schilling (2018) uses a storytelling approach to highlight traits, foibles, and ingenuity in breakthrough innovations. Personalities profiled in the book include Albert Einstein, Elon Musk, Nikola Tesla, Marie Curie, Thomas Edison, and Steve Jobs. Lockwood and Papke (2018) cover the deliberate pathways for accomplishing innovation through personal dedication. Mehta (2017) uses the "biome," a large naturally occurring community of flora and fauna occupying a major habitat (e.g., forest or tundra) to illustrate how a fertile environment can be created to facilitate a natural occurrence of innovation. Essentially, his hypothesis is the creation of a business environment, where innovation can occur and thrive. Verganti (2009) highlights the importance of design in facilitating competition, which drives innovation. A cultural comparative study of military innovation in Russia, the United States, and Israel is the focus of Adamsky (2010). He studied the extent of different strategic cultures on the approaches to military innovation in the three countries. Lessons learned from each culture can influence innovation responses in the other countries. There is an art to innovation, as opined by Kelley (2016). Most of the case examples described in the book point to the need to consider human factors and ergonomics in the pursuit of sustainable innovation. On the flip side of the art are the myths of innovation funnily described by Berkun (2010). Grissom et al. (2016) provide six pieces of evidence of innovation in the US Air Force. The literature review confirms that innovation is not new to the US Air Force and has been practiced since the official birthday of the US Air Force on September 18, 1947. Humans have pursued and actualized innovation for centuries. What is different now that suddenly makes innovation a hot topic in today's operational climate? The conjecture is that the increasingly complex and global interfaces of our current civilization call for new ways of doing things. Thus, innovation is simply a new way of doing things. Things that have been assumed and done by default in the past now require new looks and innovative approaches. That means that innovation has "the need for change" as its causal foundation. Consider the air travel security-centric changes that have occurred since September 11, 2001 (aka 9/11). The changes are innovative and responsive to the threats of today. Anyone or organization who is ready and receptive to change is essentially embracing, practicing, and actualizing innovation, which will lead to achieving the benefits of innovation. On this basis, the methodology of this chapter centers on modeling a change-focused environment to facilitate innovation.

When we talk of innovation, we often focus only on the technological output of the effort. But most often than not, innovation is predicated on the soft side of the enterprise, including people and process. The technical side of innovation cannot happen unless the people and process sides are adequately included. Thus, a systems theoretic approach is essential to realizing the goals of innovation. Everything about innovation is predicated on a systems view of the mission environment. A system is often defined as a collection of inter-related elements who collective output is higher than the sum of the individual outputs. For the purpose of innovation, a system is a group of objects joined together by some regular interaction or interdependence toward accomplishing some purpose. An innovation system must be delineated in terms of a system boundary and the system environment. This means that all organizational assets imping upon the overall output. The resources applied to the progression of innovation consist of three organizational assets, namely,

1. People
2. Technology
3. Process.

The efficient and effective application of these assets is what generates the desired output of innovation. The quantitative methodology presented in this chapter focuses on the people aspect of innovation. Specifically, we consider the learning curve implication of people in an innovation environment.

Related quantitative methodologies can be developed for technology management and process development.

An innovation system may be a team or organization, consisting of many elements, which interrelate and interact with one another and with the environment within which the system operates. The health and well-being of the whole system depends on the health and well-being of all the interrelated and interacting elements (particularly people) and the effectiveness of its responsiveness to the challenges in its environment. In the context of innovation, a system involves the interactions between people, technology, and process. Systems thinking is a mindset, which applies the systems approach to analyze and synthesize an organization's operations with the objective of resolving system deficiencies. The core of systems thinking rests in the ability to discern patterns that adequately describes the organization and the people within it. For example, typical questions related to agility and innovation in a complex technology acquisition environment may be the following:

- What is management's experience with the agile principles? What are the priorities? Are the priorities known and accepted by everyone?
- What is the experience with having all the right individuals in the program for innovative improvements?
- Are the correct metrics in place for assessing the outputs of innovation? Have they changed?
- What innovative methodologies are being employed and where?

Umbrella Theory for Innovation

Extensive literature review concludes that an overarching theory was lacking to guide the process of innovation. The key to a successful actualization of innovation centers on how people work and behave in team collaborations. Hence, the proposed methodology of "Umbrella Theory for Innovation" takes into account the interplay between people, tools, and process. The theory is illustrated in Figure 4.2. The Umbrella Theory for Innovation capitalizes on the trifecta of human factors, process design, and technology tool availability within the innovation environment. The theory harnesses the proven efficacies of existing tools and principles of systems engineering and management. This chapter selected two specific options for this purpose, the Triple C principles of project management (Badiru, 2008) and the design, evaluation, justification, and integration (DEJI) model for systems engineering processes (2014, 2019).

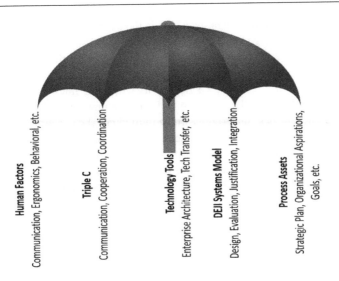

FIGURE 4.2   Umbrella Theory for Innovation. (Adapted from Badiru, 2019.)

A semantic network, also called a frame network, is a knowledge base that represents semantic relationships between elements in an operational network or system. It is often used for knowledge representation purposes in software systems. In innovation, a semantic network can be used to represent the relationships among elements (people, technology, and process) in the innovation system. This representation can give a visual cue of the critical paths in the innovation network. Postulations about the requirements for the success of an innovation effort include the following:

1. *Relative advantage:* This is the degree to which an innovation is perceived as better than the idea it supersedes by a particular group of users, measured in terms that matter to those users, like economic advantage, social prestige, convenience, or satisfaction. The greater the perceived relative advantage of an innovation, the more rapid its rate of adoption is likely to be. There are no absolute rules for what constitutes "relative advantage." It depends on the particular perceptions and needs of the user group.

2. *Compatibility with existing values and practices:* This is the degree to which an innovation is perceived as being consistent with the values, past experiences, and needs of potential adopters. An idea that is incompatible with their values, norms, or practices will not be adopted as rapidly as an innovation that is compatible.

3. *Simplicity and ease of use:* This is the degree to which an innovation is perceived as difficult to understand and use. New ideas that are simpler to understand are adopted more rapidly than innovations that require the adopter to develop new skills and understandings.
4. *Trialability:* This is the degree to which an innovation can be experimented with on a limited basis. An innovation that is triable represents less uncertainty to the individual who is considering it.
5. *Observable results:* The easier it is for individuals to see the results of an innovation, the more likely they are to adopt it. Visible results lower uncertainty and also stimulate peer discussion of a new idea, as friends and neighbors of an adopter often request information about it.

INNOVATION READINESS MEASURE

Badiru (2019) presented the framework for an innovation assessment tool. The tool is designed to assess the readiness of an organization on the basis of desired requirements with respect to pertinent factors.

Based on the spread of innovation requirements over the relevant factors, a quantitative measure of the innovation readiness of the organization can be formulated as follows:

Assuming that each checkmark can be rated on a scale of 0–10, the following composite measure can be derived:

$$IR = \sum_{i=1}^{N} \sum_{j=1}^{M} r_{ij},$$

where:

IR = innovation readiness of the organization
N = number of requirements
M = number of factors
r_{ij} = alignment measure of requirement i with respect to factor j.

The above measure can be normalized on a scale of 0–100, on the basis of which organizations and/or units within an organization can be compared and assessed for innovation readiness. Obviously, an organization that is competent in executing and actualizing innovation will yield a higher innovation readiness measure.

INNOVATION RISK MANAGEMENT

Risk is an essential element of innovation. No risk means no accomplishment. The important thing is to manage risk constructively. Risk management is an integral part of innovation. For innovation, particularly for those dealing with new ventures, risk management can be carried out effectively by investigating and identifying the sources of risks associated with each activity. These risks can be assessed or measured in terms of likelihood and impact. Because of the exploration basis of new technology, a different and diverse set of risk concerns will be involved. So, as risks are assessed for managerial processes, technical and managerial risks must also be assessed. The major activities in innovation analysis consist of feasibility studies, design, transportation, utility, survey works, construction, permanent structure works, mechanical and electrical installations, maintenance, and so on.

Definition of Risk

Risk is often ambiguously defined as a measure of the probability, level of severity, and exposure to all hazards for a project activity. Practitioners and researchers often debate the exact definition, meaning, and implications of risk. Two alternate definitions of risk are presented below:

> Risk is an uncertain event or condition that, if it occurs, has a positive or negative effect on a project objective.
> Risk is an uncertain event or set of circumstances that, should it occur, will have an effect on the achievement of the project's objectives.

In this book, we present the following definition of risk management:

> Risk management is the state of having a contingency ready to respond to the impact (good or bad) of occurrence of risk, such that risk mitigation or risk exploitation becomes an intrinsic part of the project plan.

For any innovation undertaking, there is always a chance that things will not turn out exactly as planned. Thus, project risk pertains to the probability of uncertainties of the technical, schedule, and cost outcomes of the project. All technology-based projects are complex, and they involve risks in all the phases of the project starting from the feasibility phase to the operational phase. These risks have a direct impact on the project schedule, cost, and performance.

These projects are inherently complex and volatile with many variables. A proper risk mitigation plan, if developed for identified risks, would ensure better and smoother achievement of project goals within the specified time, cost, and technical requirements. Conventional project management techniques, without a risk management component, are not sufficient to ensure time, cost, and quality achievement of a large-scale project, which may be mainly due to changes in scope and design, changes in government policies and regulations, changes in industry agreement, unforeseen inflation, underestimation, and improper estimation. Projects, which are exposed to such risks and uncertainty, can be effectively managed with the incorporation of risk management throughout the projects' life cycle.

Sources of Uncertainty

Project risks originate from the uncertainty that is present in all projects to one extent or another. A common area of uncertainty is the size of project parameters, such as time, cost, and quality with respect to the expectations of the project. For example, we may not know precisely how much time and effort will be required to complete a particular task. Possible sources of uncertainty include the following:

> Poor estimates of time and cost
> Lack of a clear specification of project requirements
> Ambiguous guidelines about managerial processes
> Lack of knowledge of the number and types of factors influencing the project
> Lack of knowledge about the interdependencies among activities in the project
> Unknown events within the project environment
> Variability in project design and logistics
> Project scope changes
> Varying direction of objectives and priorities.

IMPACTS OF REGULATIONS

Risks can be mitigated, not eliminated. In fact, risk is the essence of any enterprise. In spite of government regulations designed to reduce accident risks, accidents will occasionally happen. Government regulators can work with organizations to monitor data and operations. This will only preempt a fraction of potential risks of incidents. For this reason, government agencies must work with organizations to ensure that adequate precautions are taken in all operating scenarios. Government and industry must work together in a risk mitigation partnership, rather than in an adversarial and dictatorial relationship. There is no risk-free activity in business and industry of today. For example, many of the safety and security incidents observed over the years involved human elements – errors, incompetence, negligence, and so on. How do you prevent negligence? You can encourage nonnegligent operation or incentivize perfect record, but human will still be human when bad things happen. Effective risk management requires a reliable risk analysis technique. Below is how to deal with risk management:

> Avoid
> Assign
> Assume
> Mitigate
> Manage.

Below is a four-step process of managing risk:

> STEP ONE – Identify the risks
> STEP TWO – Assess the risks
> STEP THREE – Plan risk mitigation
> STEP FOUR – Communicate risk

We must venture out on the risk limb in order to benefit from what the innovation offers. Many leaders profess the call of "taking risk," but guidance is often lacking on to what extent risks can be taken. A quote that typifies the benefit of taking risk is echoed below:

> Behold the lowly Turtle - he only makes
> progress when he sticks his neck out.
>
> James Conan Bryan

Let us take another look at the basic definition of risk:

> Risk – "Potential Realization of an Unwanted Negative Consequence"
> Reward – "Potential Realization of a Desired Positive Consequence"

A master list of risk management involves the following:

- New technology
- Functional complexity
- New versus replacement
- Leverage on company
- Intensity of business need
- Interface existing applications
- Staff availability
- Commitment of team
- Team morale
- Applications knowledge
- Client information systems knowledge
- Technical skills availability
- Staff conflicts
- Quality of information available
- Dependability on other projects
- Conversion difficulty
- End-date dictate
- Conflict resolution mechanism
- Continued budget availability
- Project standards used
- Large/small project
- Size of team
- Geographic dispersion
- Reliability of personnel
- Availability of support organization
- Availability of champion
- Vulnerability to change
- Stability of business area
- Organizational impact
- Tight time frame
- Turnover of key people
- Change budget accepted
- Change process accepted
- Level of client commitment
- Client attitude toward IS
- Readiness for takeover
- Client design participation
- Client participation in acceptance test
- Client proximity to IS
- Acceptance process.

Possible risk response planning can follow the following options:

Accept – Do nothing because the cost to fix is more expensive than the expected loss

Avoid – Elect not to do part of the project associated with the risk

Contingency planning – Frame plans to deal with risk consequence and monitor risk regularly (identify trigger points)

Mitigate – Reduce either the probability of occurrence, the loss, or both

Transfer – Outsource.

The perspectives and guidance offered by the Umbrella Model for Innovation Management can create an avenue for managing, controlling, or mitigating risk in innovation pursuits.

SUMMARY AND RECOMMENDATION

It is expected that the methodology of the Umbrella Theory of Innovation will inspire new research inquiries into the human factors of driving innovation in organizations. The theory has the benefit of providing coverage for all the typical nuances (qualitative and quantitative) that may be encountered in the innovation environment. Of particular importance is the consideration of the people factors of innovation. The common flawed view of innovation is that it is predicated on the acquisition of technological items. While technology may be the underpinning of a specific innovation project, more often than not, the human factors will determine the success or failure of innovation. The umbrella theory explicitly calls out the people aspects of the pursuit of innovation. The quantitative measure of innovation readiness can be adapted and expanded to fit specific research, development, and implementation themes related to the pursuit of innovation.

REFERENCES

Adamsky, D. (2010). *The Culture of Military Innovation: The Impact of Cultural Factors on the Revolution in Military Affairs in Russia, the US, and Israel.* Stanford, CA: Stanford University Press.

Badiru, A.B. (2014). Quality insights: The DEJI model for quality design, evaluation, justification, and integration, *International Journal of Quality Engineering and Technology*, 4(4), 369–378.

Badiru, A.B. (2008), *Triple C Model of Project Management: Communication, Cooperation, and Coordination*. Boca Raton, FL: CRC Press/Taylor & Francis Group.

Badiru, A.B. (2019). Quality insights: Umbrella theory for innovation. *International Journal of Quality Engineering and Technology*, 8(2), 95–102.

Berkun, S. (2010). *The Myths of Innovation*. Sebastopol, CA: O'Reilly Media, Inc.

Degraff, J. and Degraff, S. (2017). *The Innovation Code: The Creative Power of Constructive Conflict*. Oakland, CA: Berrett-Koehler Publishers, Inc.

Grissom, A.R., Caitlin, L., and Mueller, K.P. (2016). *Innovation in the United States Air Force: Evidence from Six Cases*. Los Angeles, CA: RAND.

Hamel, G. (2012). *What Matters Now: How to win in a World of Relentless Change, Ferocious Competition, and Unstoppable Innovation*. San Francisco, CA: Jossey-Bass.

Kelley, T. (2016). *The Art of Innovation*. London: Profile Books Ltd.

Keeley, L., Pikkel, R., Quinn, B., and Walters, H. (2013). *Ten Types of Innovation: The Discipline of Building Breakthroughs*. Hoboken, NJ: John Wiley & Sons.

Lockwood, T. and Papke, E. (2018). *Innovation by Design: How Any Organization Can Leverage Design Thinking to Produce Change, Drive New Ideas, and Deliver Meaningful Solutions*. Newburyport, MA: Career Press; Santa Monica, CA: RAND Corporation.

Mehta, K. (2017). *The Innovation Biome*. Austin, TX: River Grove Books.

Satell, G. (2017). *Mapping Innovation: A Playbook for Navigating a Disruptive Age*. New York: McGraw-Hill.

Schilling, M.A. (2018). *Quirky. The Remarkable Story of the Traits, Foibles, and Genius of Breakthrough Innovators Who Changed the World*. New York: Public Affairs, Hachette Book Group.

Verganti, R. (2009). *Design-Driven Innovation: Changing the Rules of Competition by Radically Innovating What Things Mean*. Boston, MA: Harvard Business School Publishing.

Voehl, F., Harrington, H.J., Fernandez, R., and Trusko, B. (2019). *The Framework for Innovation: A Guide to the Body of Innovation Knowledge*. Boca Raton, FL: CRC Press/Taylor & Francis Group.

Leadership Requirements for Innovation

5

INTRODUCTION

Innovation can mean different things to different people. This chapter is based on McCauley (2017) and presents conceptual and operational requirements for leadership in innovation. An organization can be innovative in process-improvement strategies without being involved in technological innovation. Foremost in process-improvement innovation is the role of organizational leadership. Good and consistent leadership is essential for facilitating support, allocating resources, and sustaining motivation for innovation throughout the organization.

Transformative thinking is required for enhancing innovation in both small and large organizations. The larger an organization, the more coordinated transformation will be required, particularly from the standpoint of education. National thought leaders and organizations such as the National Academy of Engineering are supporting projects to explore this relationship. The Educate to Innovate (ETI) project was designed to explore the issue regarding teaching innovation and the expected outcome, entrepreneurship.

During the 1950s and 1960s, Sputnik and the space race stimulated a generation of Americans to follow education and careers in science and technology (S&T). Half a century later, American students are now graded 22nd and 21st among their peers all over the world in science and math, respectively. Students in the United States, formerly a leader in science, technology, engineering, and mathematics (STEM), are now outperformed by students from Slovenia, Hungary, and Estonia, among others (McCauley, 2017).

In 1983, the National Commission on Excellence in Education published *A Nation at Risk*, a nationwide study that highlighted the intolerable state of the American education system:

> Our nation is at risk. Our once unchallenged preeminence in commerce, industry, science, and techno logical innovation is being overtaken by competitors throughout the world. This report is concerned with only one of the many causes and dimensions of the problem, but it is the one that undergirds American prosperity, security, and civility. What was unimaginable a generation ago has begun to occur-others are matching and surpassing our educational attainments. If an unfriendly foreign power had attempted to impose on America the mediocre educational performance that exists today, we might well have viewed it as an act of war.

More than two decades afterward, in 2010, the National Academies of Science, Engineering, and Medicine published *Rising above the Gathering Storm, Revisited: Rapidly Approaching Category 5*, which built on the findings of its 2005 *Gathering Storm* report. Notably, the report warns that

> Today, for the first time in history, America's younger generation is less well-educated than its parents.

In an effort to respond to the faltering academic status of American students and in a quest to elevate them "from the middle to the top of the pack in science and math," the Obama Administration announced its ETI initiative in November 2009.

President Barack Obama's ETI campaign is publicized as a joint effort between the federal government, the private sector, and the nonprofit and research communities to raise the standing of American students in science and math through dedication of time and money and volunteering. The program attempts to enhance STEM literacy, improve teaching quality, and develop educational and career opportunities for America's youth.

At the time the program was first declared in November 2009, the participating organizations offered a financial and in-kind commitment of more than $260 million. Taxpayer commitments for the federal government's portion of ETI add to that total. In addition, five public–private partnerships were announced, as well as commitments by key societal and private sector leaders to muster funds for STEM education, innovation, and awareness. These partnerships and commitments are as follows:

- Time Warner Cable's "Connect a Million Minds" (CAMM), which pledges to connect children to after-school STEM programs and activities in their area.

- Discovery Communications' "Be the Future" will broadcast dedicated science programming to more than 99 million homes and offer interactive science education to approximately 60,000 schools.
- Sesame Street's "Early STEM Literacy" commits to a 2-year focus on STEM subjects.
- National Lab Day will promote hands-on learning with 100,000 teachers and 10 million students over the next 4 years and foster communities of collaboration between volunteers, students, and educators in STEM education. These initiatives will then culminate in a nationally recognized day centered on science activities.
- The National STEM Video Game Challenge promotes the design and creation of STEM-related video games.
- The annual White House Science Fair will bring the winners of science fairs from across the nation to the White House to showcase their STEM creations and innovation.

Sally Ride, the first female astronaut, Craig Barrett, the former Intel chairman, Ursula Burns, CEO of Xerox, and Glenn Britt, CEO of Eastman Kodak committed to foster interest and support for STEM: education among American corporations and philanthropists.

In January 2010, President Obama announced the continuation of the program, stressing the half-billion-dollar monetary obligation from the administration's partners. This development includes an additional commitment of $250 million in financial and in-kind support and a pledge by 75 of the nation's biggest public universities to train 10,000 new teachers by 2015. The program expansion also incorporated additional public–private partnerships anticipated to aid the training of new STEM educators, together with the launch of Intel's Science and Math Teachers Initiative and the PBS Innovative Educators Challenge, as well as the expansion of the National Math and Science Initiative's UTeach program and Woodrow Wilson Teaching Fellowships in math and science. In addition, the president called on 200,000 federal government staff working in the fields of S&E to volunteer to work with educators in order to foster enhanced STEM education.

STEM-educated workforce is very important for the protection and the wealth of the United States as industry and government increasingly demand exceedingly trained STEM professionals to vie in the international market and look to S&T to help stay one step ahead of national security threats. The United States must not permit itself to be outcompeted in STEM. While the administration's ETI enterprise is projected to raise the United States "from the middle to the top of the pack in science and math," this one-size-fits-all federal approach fails to cure the primary problems of educational performance and does not stop the permeable pipeline in the American education system.

EVOLUTION OF INNOVATION

The principles associated with innovation can be applied to organizations, individuals, and product development. These three categories of innovation can also be applied simultaneously to create a culture, where individuals are continually seeking to be innovative and create enhanced product outcomes. The meaning of innovation as evolved with US Federal funding agencies as well. For example, consider the National Science Foundation (NSF), one of the premier research funding agencies in the United States that funds 24% of all federally supported basic research conducted by colleges and universities in the United States each year.

For many years, NSF largely focused on funding only basic research rather than funding applied research and technology transition. Now the NSF's funding goals are extending beyond basic research to support various aspects of groundbreaking applied research and the transition of research outcomes into useful products, services, and technologies. There's a good reason for this change in focus. Historically, it was thought that it could take up to 50 years for the knowledge learned from basic research to be applied to products and services. However, as the pace of change itself continues to increase, the speed of technology and new development has compressed the time it takes to move basic research from reaction to knowledge to actual application. The NSF reflects this shift quite powerfully in its want to now fund more applied research. The quick transition of the NSF's innovation core and its desire to swiftly convert new knowledge into new products and services is solid evidence of change.

I-CORP AT NSF

America's affluence grew in part from the capability to profit economically on groundbreaking developments from science and engineering research. At the same time, a well-informed, imaginative labor force has maintained the country's international leadership in significant areas of technology. These essential discoveries and competent labor force resulted from substantial, incessant investment in science and engineering. A strong capability for leveraging essential science discoveries into influential engines of innovation is necessary to maintain our competitive edge in the future. The NSF supports fundamental research and education in science and engineering. NSF's dual role, distinctive among government agencies, results in new knowledge and paraphernalia as well as a competent groundbreaking workforce. These corresponding building blocks of innovation have led to innovatory high-tech advances and completely new industries. Through this program, NSF seeks to hasten the improvement of new technologies, products, and processes that arise from elementary study. NSF investments will advantageously strengthen the innovation ecosystem by addressing the challenge inbuilt in the early stages of the innovation process. This solicitation will support partnerships that are designed to triumph over scores of obstacles in the path of innovation.

PROGRAM DESCRIPTION

The objectives of this program are to encourage translation of fundamental research, to facilitate collaboration between the academic world and business, and to train students to comprehend innovation and entrepreneurship. The rationale of the NSF I-Corps program is to spot NSF-funded researchers who will obtain extra support – in the form of mentoring and funding – to hasten the conversion of knowledge derived from essential research into up-and-coming products and services that can attract successive third-party funding.

The NSF is an autonomous federal agency created by the National Science Foundation Act of 1950, as amended (42 USC 1861–75). The act states the function of the NSF is "to promote the progress of science; and to advance the national health, prosperity, and welfare by supporting research and education in all fields of science and engineering." NSF funds research and learning in most fields of science and engineering through grants and cooperative agreements to more than 2,000 colleges, universities, K-12 school systems, businesses, informal science organizations, and other research organizations all over the United States. The foundation accounts for about one-fourth of federal support to educational institutions for essential research. NSF receives about 40,000 proposals each year for study, learning, and training projects, of which roughly 11,000 are funded. In addition, the foundation receives thousands of applications for graduate and postdoctoral fellowships. The agency operates no laboratories itself but does support national research centers, user facilities, certain oceanographic vessels, and Arctic and Antarctic research stations. The foundation furthermore supports joint research between universities and industry, US participation in global scientific and engineering efforts, and educational activities at every academic level.

The role of creativity and innovation has changed our nation because now we are pushing more to see these new developments converted into new products and services, and the driving factor in accomplishing this is leadership. There is even more accountability in terms of wanting to understand what has been done with research funding for over the past several years. Generally, Americans convey extremely favorable attitudes toward S&T. In 2001, overpowering majorities of NSF survey respondents agreed with the following statements:

- "Science and technology are making our lives healthier, easier, and more comfortable" (86% agreed and 11% disagreed).
- "Most scientists want to work on things that will make life better for the average person" (89% agreed and 9% disagreed).
- "With the application of science and technology, work will become more interesting" (72% agreed and 23% disagreed).
- "Because of science and technology, there will be more opportunities for the next generation" (85% agreed and 14% disagreed).

In addition, Americans give the impression to have more positive attitudes toward S&T than their counterparts in the United Kingdom and Japan.

Despite these positive indicators, a sizable segment, although not a majority, of the public has some reservations concerning science, especially technology. For example, in 2001, approximately 50 percent of NSF survey respondents agreed with the following statement: "We depend too much on science and not enough on faith" (46% disagreed). In addition, 38% agreed with the statement: "Science makes our way of life change too fast" (59% disagreed).

FUNDING OF INNOVATIVE RESEARCH

All indicators point to general support for government funding of essential research. In 2001, 81% of NSF survey respondents agreed with the following statement: "Even if it brings no immediate benefits, scientific research that advances the frontiers of knowledge is necessary and should be supported by the Federal Government." The level of agreement with this statement has consistently been in the 80% range. In 2000, 72% of UK residents agreed with the statement, as did 80% of Japanese residents in 1995.

These differences in the measure of public support worldwide for basic research are notable. This may be attributed to the increased expectations in terms of transitioning science to technology and innovations. The result is that people expect basic research to more readily provide benefits to society and, in fact, in 2001, 16% disagreed with the statement completely. This suggests that we can expect immediate benefits from basic research and this trend of expectation has continued.

Although there is strong evidence that the public supports the government's investment in basic research, few Americans are able to name the two agencies that provide most of the federal funds for this type of research. In a recent survey, only 5% identified the National Institutes of Health (NIH) as the agency that "funds most of the taxpayer-supported medical research performed in the United States," and only 3% named NSF as "the government agency that funds most of the basic research and educational programming in the sciences, mathematics, and engineering."

In addition, those with more positive attitudes toward S&T were more likely to express support for government funding of basic research. In 2001, 93% of those who scored 75 or higher on the Index of Scientific Promise agreed that the federal government should fund basic scientific research compared with only 68% of those with relatively low index scores.

In 2001, only 14% of NSF survey respondents thought the government was spending too much on scientific research; 36% thought the government was not spending enough, a percentage that has grown steadily since 1990, when 30% chose that answer. Men are more than likely than women to say the government is spending too little in support of scientific research (40% versus 33% in 2001).

To put the response to this item in perspective, at least 65% of those surveyed thought the government was not spending enough on other programs, including programs to improve health care, help senior citizens, improve education, and reduce pollution. Only the issues of space exploration and national defense received less support for increased spending than scientific research.

In 2001, 48% of those surveyed thought spending on space exploration was excessive, the highest percentage for any item in the survey and nearly double the number of those who felt that the government was spending too much on national defense. In contrast, the latter has been falling steadily, from 40% in 1990 to 25% in 2001.

DIVERSE VIEWS OF INNOVATION

Definitions of innovation differ, but the general thread among these definitions is that innovations present a new or better product, service, or resource that adds "value" to those seeking it. The ETI study conducted 60 interviews that revealed common characteristics of innovators. A prevailing aspect of innovation is team interaction or team activities. For these teams to be effective, they are often managed by a technical person with detailed knowledge of the proposed innovation. In these situations, it is imperative that the team leader understands how to inspire, motivate, and lead the team as they move toward a useful innovation. When innovators were asked to describe characteristics of innovations or innovative products, the following characteristics emerged:

- Innovation provides societal value
- Innovation is an improvement
- Innovation occurs at the interfaces of different functions
- Teamwork is important to the process of innovation
- Innovation is the effect of joint effort
- Innovation is part of an invention-value continuum.

TYPES OF INNOVATION

Innovation applications are commonly applied to either a product, a process, or a service. To additionally comprehend how this is done, let's reflect on three categories of innovation.

Product Innovation

Product innovation is about making valuable changes to material products. Interrelated terms that are frequently used interchangeably comprise product design, research and development, and new product development (NPD). All of these terms proffer a particular viewpoint on the degree of alteration to products. Well-known organizations characteristically have a collection of products that must be incrementally enhanced or adjusted as problems are recognized in service or as new requirements emerge. It is imperative that they also work on add-ons to the product families. One of the major actions of the product design team is the work it carries out on next-generation products or new models of products. They might also work on designing far-reaching new products or new core products that enlarge the portfolio considerably and frequently involve drastically new processes to produce them. These new core products idyllically present the organization the possibility of major increases in revenue and growth, which can also create the potential of short-term monopoly in the market.

The product development process for next-generation and new core products, according to Cooper, follows a familiar cycle in most organizations:

a. Ideation
b. Preliminary investigation
c. Detailed investigation
d. Development
e. Testing and validation
f. Market launch and full production.

All of these steps involve communication with customers, who might take part in idea creation and element recognition. Key performance criteria in the design process revolve around the following:

a. Time to market
b. Product cost
c. Customer benefit delivery
d. Development costs.

These standards can be traded off against one another. For instance, development costs can be traded against time to market, customer benefits can be traded against product costs, and so on. Three blueprint systems have ascertained themselves as providing a management system for efficient product innovation: phase review, stage gate, and product and cycle-time excellence (PACE).

a. Phase Review
This technique splits the product development life cycle into a sequence of different phases. Every phase encompasses a body of work that, once finished and evaluated, is dispensed over to the next phase. No consideration is paid to what may or may not occur in the succeeding phases, principally for the lack of knowledge or exclusive focus on the job in the existing phase. The phase review technique is a chronological rather than a simultaneous product design method, that is, each phase is accomplished and concluded before the commencement of the next phase.
b. Stage Gate
This technique is a simultaneous product design procedure that follows a prearranged life cycle from idea creation to market commencement. The stages in this technique are first and foremost cross-functional. Stage gates appear at the end of each stage, where a design evaluation takes place. Each stage gate evaluates the

decided deliverables for completion at the conclusion of the stage, a checklist of the standard agreed for each stage, and a choice about how to advance from a particular stage.

c. PACE

This method is concerned mainly with enhancing product improvement strategies. The technique connects product strategy with the general strategy and goal of the organization. A key element is positioning of the voice of the customer all through the product design procedure. Strategies are divided into six product strategic thrusts: expansion, innovation, strategic balance, platform strategy, product line strategy, and competitive strategy. Product innovation methods and processes are one element in an organization's mission to create value for customers.

Process Innovation

Process innovation can be observed as the launching of a new or considerably enhanced method for the construction or delivery of production that append value to the organization. The term "process" refers to an interconnected set of actions designed to convert inputs into a specific result for the customer. It implies a strong prominence on how work is done within an organization rather than what an organization does.

Processes recount every operational action by which value is presented to the end client, such as the purchase of raw materials, production, logistics, and after-sales service. The process innovation in the 1970s and 1980s gave Japanese manufacturing a viable advantage that permitted them to take over some international markets with cars and electronic goods. Likewise, process innovation has permitted organizations such as Dell and Zara to achieve competitive advantage by offering higher-quality products, delivered faster and more proficiently to the market than by the competitors. By focusing on the resources by which they transform inputs, such as raw materials, into results, such as products, organizations have achieved efficiencies and have added importance to their production. Process innovation permits some organizations to contend by having a further proficient value chain than their rivals have.

Process innovation has resulted in organizational enhancement such as lower stock levels; quicker, additional flexible production processes; and more responsive logistics. Organizations can develop the competence and value of their processes with a huge array of diverse enablers. Even though the use of these enablers is dependent on the organizational framework, many present the possibilities for improved process performance. The application of technology such as robotics, enterprise resource planning systems, and sensor technologies can change the process by decreasing the price or variation of its output, improving safety, or decreasing the throughput time of the process.

Service Innovation

Service innovation is concerned with making changes to intangible products. Services are frequently linked with work, play, and recreation. Examples of these types of service consist of education, banking, government, recreation, entertainment, hospitals, and retail stores. In the past decade, an enormous amount of knowledge-based services has been accessible through websites. These services involve intangible products, have a high quantity of customer dealings, and are typically set in motion on demand by the customer. Defining a service can be to some extent problematic. Some define service as a sequence of overlapping value-creating activities.

Others define service in terms of performance, where customer and provider coproduce value. There are three categories of service operations:

a. Quasi-manufacturing (e.g., warehouses, testing labs, recycling)
b. Mixed services (e.g., banks, insurance, realtors)
c. Pure services (e.g., hospitals, schools, retail).

Services can without a doubt involve products that form a comprehensive part of the product life cycle, from preliminary sales to end-of-life recycling and clearance. Service business in areas such as finance, food, education, transportation, health, and government make up most organizations in any economy.

These organizations as well require innovation incessantly so that they can enhance levels of service to their customers. A key characteristic of a service is a very high level of communication with the end user or customer. The customer is often not capable of separating the service from the person delivering the service and so will make quality postulation based on impressions of the service, the group delivering the service, and any product delivered as part of the service. An additional feature of some service organizations is that their product may be perishable; consequently, the product must be consumed as soon as possible following purchase. Consequently, the timing of the delivery and customer opinion of quality are vital to success.

The notion of service quality is of particular significance. Service quality is a function of numerous factors including the uniqueness of offerings, intangibilities such as customized customer contact or perishable manufacture, and a continued capacity for innovations of the service. Another important driver of service innovation comes from the possibilities afforded by the new information technology podium, predominantly the Internet. The Internet is a priceless resource on which new service associations between organizations and their customers are being developed every day.

INNOVATION AND ENTREPRENEURSHIP

If innovation is successful, the expected outcome is the transitioning of these new products, processes, or services into useful products that people are willing to pay for in the United States and globally. Although innovation and entrepreneurship are related, many caution the intent of focusing too much on entrepreneurship in the initial stages of the creative aspect of innovation. This perspective believes that entrepreneurship should be a natural outcome of entrepreneurship but should not be the initial focus.

It is really important to lead with "innovation" and have it evolve into "entrepreneurship" because innovation is the large end of the funnel that appeals to and actually requires participation by a much broader audience (McCauley, 2017). Nonbusiness, non-engineering, and non-STEM people are every bit as important to include in that innovation process because the process is not as rich and has inferior outcomes without that diversity. In order to see this type of innovation systematically realized, engineering leaders must understand principles that should be integrated into the creative process to produce effective innovations.

The terms "entrepreneurship" and "innovation" are used interchangeably over and over again, nevertheless this is deceptive. Innovation is frequently the starting point on which an entrepreneurial business is built for the reason of the competitive advantage it offers. On the contrary, the act of entrepreneurship is simply one means of bringing an innovation result to the marketplace. Technology entrepreneurs regularly decide to build a startup company for a technological innovation. This will offer financial and skill-based resources that will take advantage of the chance to grow and commercialize the innovation. Once the entrepreneur has set up a business, the focal point shifts in the direction of its sustainability, and the best way to attain this is through managerial innovation. Nonetheless, innovation can be conveyed to the market by ways other than entrepreneurial startups; it can also be subjugated through well-known organizations and deliberate alliances between organizations.

INNOVATION CASE STUDY: CHARLES DOW

In 1896, Charles Dow created the Dow Jones Industrial Average in order to provide a snapshot of the US economy through the stock market. There were 12 companies on Dow's original list American Cotton Oil, American Sugar, American Tobacco, Chicago Gas, Distilling & Cattle Feeding, General Electric (GE), Laclede Gas, National Lead, North American, Tennessee Coal and Iron, US Leather pfd., and US Rubber. Of all of those companies, which were financial leaders at the turn of the 20th century, there is only one you might recognize that is still in business today: General Electric.

What is the key to GE's century-long tenure? Product innovation. According to business researchers, Heath Downie and Adela J. McMurray, "The consistency of GE's commitment to product innovation was made possible by the steadiness of the company's leadership." Even during the Great Depression, GE found a way to allocate diminishing financial resources to its research and development initiatives.

Today, GE has taken its commitment to innovation even further, crowdsourcing both internally and externally to drive advancements in several industries. In fact, GE has an Open Innovation Manifesto, which states the following:

> We believe openness leads to inventiveness and usefulness. We also believe it's impossible for any organization to have all the best ideas, and we strive to collaborate with experts and entrepreneurs everywhere who share our passion to solve some of the world's most pressing issues. We'll never stop experimenting, collaborating and learning we'll get smarter as we go, and the Global Brain will evolve and grow with us.

GE has a hand in advancing just about every engineering industry you can think of such as aviation, software, consumer goods, water and wastewater, power and energy, transportation, and health care, to name a few. Named "America's Most Admired Company" in a poll conducted by *Fortune* magazine and one of "The World's Most Respected Companies" in polls by *Barron's* and the *Financial Times*, the quality work GE has done for the planet has not gone unnoticed, and the company's leadership is extremely dedicated to quality and innovation. GE's Ecomagination is a business initiative designed to develop innovative solutions to environmental challenges while driving economic growth. Innovation is the foundation for Ecomagination, and the company has developed a lot of solutions that solve complex problems for a multitude of industries. Ecomagination has really been the catalyst within

GE to step outside and get those ideas and that outside innovation moving forward internally.

In 2016, GE announced it would be relocating its corporate headquarters to Boston, Massachusetts, in part to enable GE to place additional emphasis on digital industrial innovation. This is further proof of the company's commitment to innovation and its leadership push to improve access to a more innovative workforce and relocate to a better environment for innovation. In essence, leadership drives innovation.

REFERENCE

McCauley, P. (2017). *Essentials of Engineering Leadership and Innovation.* Boca Raton, FL: CRC Press/Taylor & Francis Group.

Organizational Readiness for Innovation

6

INTRODUCTION

Thal and Shahady (2019) pose the question, "Is Your Organization Ready for Innovation?" by echoing the following quote from the defense industry:

> None of the most important weapons transforming warfare in the 20th century – the airplane, tank, radar, jet engine, helicopter, electronic computer, not even the atomic bomb – owed its initial development to a doctrinal requirement or request of the military.

The opening quote from Chambers (1999) suggests that the defense community has been at the forefront of innovation over the past century. Despite their success though, many organizations in the defense community struggle to explain specifically what they do to facilitate and implement innovation. To some, "being innovative" is interpreted as a means to empower employees to make decisions and solve problems at the lowest level possible. To others, "being innovative" is viewed as having open work spaces that lead to increased collaboration. However, innovation requires a much deeper understanding if it's to be successful. Beyond acknowledging the importance of innovation and inspiring the workforce though, what can leaders do to ensure their organizations are ready for innovation? To help answer that question, we think a good place to start is to review the organization's processes and dynamic capabilities. In many ways, these two concepts represent the DNA of the organization – and whether the organization is structured to facilitate innovation. We will then introduce a conceptual model that leaders can use to foster disruptive innovation. These three concepts – processes, dynamic capabilities, and the conceptual model – are equally applicable to organizations in both the public and private sectors.

Background

From the first powered flight by the Wright brothers in 1903 and the use of airplanes in the Army Air Corps to modern-day advances in military airpower, it's often said that innovation is a part of the Air Force culture. General Henry "Hap" Arnold alluded to this in 1945 when he suggested that "… any air force which does not keep its doctrines ahead of its equipment, and its vision far into the future, can only delude the nation into a false sense of security." Innovation has subsequently been highly touted by many of the Air Force's past and present leaders as being critical to the future success of the service. Furthermore, the vision statements for many Air Force organizations also acknowledge the importance of innovation. In fact, the Air Force's current vision statement is: The World's Greatest Air Force – Powered by Airmen, Fueled by Innovation.

Despite the importance placed on innovation though, two recent studies exploring the use of experimentation in innovation reported some sobering results. In the first study, the US Air Force Scientific Advisory Board (SAB) concluded that the Air Force is very good at sustaining innovation but has "largely lost its ability to foster disruptive innovation" (USAF SAB, 2006). The SAB also concluded that Air Force organizations have not created an environment conducive to innovation. In the second study, the Air Force Studies Board (AFSB) expressed similar findings. Some of their key observations included a lack of space, time, and funding for experimentation-driven innovation, a fear of failure, a lack of appropriate processes, and a culture that is not supportive of innovation (AFSB, 2016). The results from both studies seem to indicate a stagnant environment in which the Air Force has lost momentum when it comes to technological innovation and is at risk of becoming irrelevant in the future battlespace.

Organizational Processes

As W. Edwards Deming is fond of saying, "If you can't describe what you're doing as a process, then you don't know what you're doing." Let's put this into proper context for this chapter – if organizations are unable to describe their innovation efforts as a process, they're probably struggling with being innovative. This is consistent with Drucker (2002), who states that innovation is "capable of being presented as a discipline, capable of being learned, capable of being practiced." In other words, to make innovation more successful, it helps to view it as a process – a process that can be managed.

Processes are prevalent in organizations; they can be found in the way organizations operate, in their structures and cultures, and in the mindset of senior leadership (O'Reilly and Tushman, 2007). For those who may not have given it much thought, most organizations contain three general types of processes. Primary processes, also referred to as business processes, tend to be cross-functional. They often reflect the unique competencies of the organization and provide direct value to the customer; therefore, they are often considered mission essential. Support processes, on the other hand, usually do not provide direct value to the customer; instead, they are fairly standard and help sustain the organization. Common examples include management of information technology, infrastructure, capacity, and human resources. Finally, management processes provide direction and governance to ensure that the organization operates effectively and efficiently. They are generally conducted by senior leaders to develop and deploy strategy, manage the organizational structure, and establish organizational performance goals.

Regardless of its type, any process is "an organized group of related activities that work together to transform one or more kinds of input into outputs that are of value to the customer" (Hammer and Champy, 2001). Processes are, thus, designed to achieve a specific goal – a goal that, in turn, provides value to customers (either internal or external). This implies that processes are not random or ad hoc. Furthermore, every process in an organization should be viewed as either contributing to an organization's success or adding to its bureaucratic inefficiency – the key is being able to identify those processes that are a detriment to the organization and taking action to change them. When talking about processes and bureaucracy, an old adage often found in fortune cookies comes to mind: "People will do tomorrow what they did today because that is what they did yesterday." Ed de Bono refers to this as the "continuity of time sequence." Trapped by the sequence of our experiences, processes have a habit of developing almost arbitrarily yet becoming permanent.

It's human nature – and it explains a lot. It explains why many of today's practices are a reflection of "that's the way we've always done it." It explains how redundant processes develop and add to an organization's overhead. It shows how bureaucracy grows incrementally over time. Finally, it explains why few organizations run the way they should. The problem usually isn't about competence or effort – more often than not, the processes are the problem. Consider the following excerpt from Morison (1966):

> A time-motion expert … watched one of the gun crews of five men at practice in the field for some time. Puzzled by certain aspects of the procedures, he took some slow-motion pictures … A moment before the firing, two members of the gun crew ceased all activity and came to attention

for a three-second interval extending throughout the discharge of the gun. He summoned an old colonel of artillery, showed him the pictures, and pointed out this strange behavior. What, he asked the colonel, did it mean. The colonel, too, was puzzled. He asked to see the pictures again. "Ah," he said when the performance was over, "I have it. They are holding their horses."

The description above relates to horse artillery units supporting the cavalry. However, as technology advanced and the process of firing artillery guns changed, part of the previous procedure remained intact. An argument can certainly be made regarding the importance of upholding tradition, especially in military organizations. In many other cases though, does the tradition provide value? Or is it simply a carryover from the past because that's the way it's always been done?

If we extend this line of reasoning to processes in general, how many processes in our organizations are simply carryovers from the past? To give this some critical thought, it might be helpful to evaluate the organization's core competencies. When organizations excel at an activity, they can easily become overcommitted to it. If the organization holds on to them too tightly, those core competencies and their accompanying processes can easily become core rigidities (Leonard-Barton, 1992). In our own organizations, how many similar examples exist? How many processes do we have that were built in a different era and possibly for different purposes but continue to be blindly followed? Breaking away from these processes and the past requires conscious effort – it requires the will to question the existing processes and the inherent assumptions on which those processes were based.

As we review our organization's activities and processes, an important concept to consider is the value chain, which represents the primary and limited support processes that provide value to the customer (Porter, 1985). While organizations may have hundreds of work processes, they usually have very few business processes. As such, value-creating business processes begin and end with the external customer, tend to be large in scope, and commonly span multiple organizational components. Since this group of processes represents the core competencies of the organization, this is where performance improvement work is often focused. Furthermore, these processes must be aligned and integrated to enable effective performance of the organization.

With this brief introduction to processes, the question for most organizations is whether innovation is considered a core competency. If it is, does the organization treat innovation as a process that can be managed? And do other processes within the organization align with and complement the innovation process? An approach organizations might take to address these questions is to review their capabilities.

Organizational Capabilities

When examining the success of organizations, a fundamental question that often arises is, "Why does a particular organization or group of organizations outperform other similar organizations?" To answer the question, two schools of thought have developed: the industry-based view and the resource-based view (RBV). The industry-based view assumes that success has something to do with the industry in which the organization operates; therefore, strategies are based on an external analysis (such as Porter's Five Force model). On the other hand, the RBV assumes that success has something to do with the assets (or resources) the organization own and control; therefore, strategies are based on an internal analysis. Since empirical evidence suggests that organizational differences account for more variation in performance than industry differences (Rumelt, 1984), the RBV has been increasingly referenced in the strategy literature. Although the RBV framework was initially developed to understand how businesses achieve and sustain competitive advantage (Prahalad and Hamel, 1990; Barney, 1991), its inward focus also makes it appealing to public sector organizations (Matthews and Shulman, 2005; Pablo et al., 2007).

Eisenhardt and Martin (2000) suggest that assets – physical, human, and organizational resources – are the foundation of the RBV approach. Furthermore, the "bundling" of these assets to perform specific business processes is often referred to as a capability. Organizational capabilities are, thus, the various routines (or patterns) and processes that transform inputs (i.e., resources) into outputs (i.e., goods and services that provide value to the customer). Routines represent sequences of actions for performing tasks in an organization. Institutionalized through technologies, formal procedures, and informal conventions or habits, they reflect "the way we do things around here."

Organizational capabilities can be characterized as either ordinary or dynamic as summarized below:

Ordinary capabilities (doing things right)
- Technical efficiency in basis business functions
- Operational, administrative, and governance
- Relatively easy, imitable.

Dynamic capabilities (doing the right things)
- Strategic fit over the long run (evolutionary fitness)
- Sensing, seizing, shaping, and transforming
- Difficult; inimitable.

Ordinary capabilities represent the routines and standard operating procedures within the organization. They tend to support the day-to-day operations of the organization and change little over time; in some cases, they are often referred to as "best practices." Dynamic capabilities are the real reason for an organization's long-term success; they represent a set of abilities that enable an organization to quickly build capability and affect change. Organizations with dynamic capabilities are, thus, better positioned to exploit opportunities by adapting organizational structures and routines. According to Teece (2006), strong ordinary capabilities are necessary but not sufficient for long-term success; they can be acquired (or "bought") from other organizations or through investments in training. However, strong dynamic capabilities are necessary and sufficient for long-term success; they cannot be bought and must be built. From an innovation perspective, this is a critical point – the ability to build and improve effective routines is often considered a necessary ingredient for successful innovation.

Teece et al. (1997) define dynamic capabilities as an organization's "ability to integrate, build, and reconfigure internal and external competencies to address rapidly changing environments." The term "dynamic" is meant to indicate an organization's capacity to establish new competencies in response to environmental conditions and the ability to reconfigure their assets and develop new routines (Lee and Kelley, 2008; Eisenhardt and Martin, 2000), while the term "capabilities" is meant to imply the importance of strategic management. Taken together, dynamic capabilities serve as the source of an organizations' competitive advantage. Additionally, they are often considered a necessary component of the innovation process (Lee and Kelley, 2008). To be specific, Lawson and Samson (2001) suggest three primary reasons dynamic capabilities align with innovation efforts: (1) the lack of a technology focus recognizes the importance of other resources; (2) the RBV basis makes it applicable to product, process, system, and business model innovation; and (3) asset heterogeneity reflects the expectation that there is no one generic formula. From a dynamic capability perspective, Tidd and Bessant (2009) describe the core abilities in managing innovation summarized below:

Basic ability: Contributing routines
Recognizing: Searching the environment for technical and economic clues to trigger the process of change
Aligning: Ensuring a good fit between the overall business strategy and the proposed change – not innovating because it is fashionable or as a knee-jerk response to a competitor
Acquiring: Recognizing the limitations of the company's own technology base and being able to connect to external sources of knowledge, information, equipment, etc. Transferring technology from various outside sources and connecting it to the relevant internal points in the organization

Generating: Having the ability to create some aspects of technology in-house – through R&D, internal engineering groups, etc.

Choosing: Exploring and selecting the most suitable response to the environmental triggers which fit the strategy and the internal resource base/external technology network

Executing: Managing development projects for new products or processes from initial idea through to final launch; monitoring and controlling such projects

Implementing: Managing the introduction of change – technical and otherwise – in the organization to ensure acceptance and effective use of innovation

Learning: Having the ability to evaluate and reflect upon the innovation process and identify lessons for improvement in the management routines

Developing the organization: Embedding effective routines in place – in structures, processes, underlying behaviors, etc.

Teece et al. (1997) describe organizational processes as routines of current practice serving three roles: coordination/integration, learning, and reconfiguration. These routines are used to integrate and exploit competencies. However, what the organization can accomplish with its dynamic capabilities is constrained by its asset positions and shaped by evolutionary and coevolutionary paths (Teece et al., 1997). An organization's position reflects specific competencies in both tangible and intangible assets; these competencies may consist of technology capabilities, complementary assets, external relationships, etc. Paths represent options available to organizations based on core competencies, technology trajectories, and emerging opportunities. They tell us that the availability of current strategic choices is a reflection of past strategic choices (Teece et al., 1997). In other words, it is typically difficult for most organizations to ignore what has been done in the past and develop new ideas. Specifically in a research and development (R&D) environment, Cohen and Levinthal (1990) argue that an organization's innovative capability is a function of its prior related knowledge; without prior experience, the organization would not be in a position to recognize value and exploit it.

O'Reilly and Tushman (2007) suggest that capabilities are the result of senior leader actions to facilitate and ensure learning, integration, reconfiguration, and transformation; these processes thus dictate the paths (i.e., strategic choices) organizations take. They typically refer to this as the "sensing and seizing" of new opportunities to emphasize the key role of strategic management. Other researchers also include the role of "transforming" when referring to dynamic capabilities. As organizational leaders ponder their role, and the actions they take, to facilitate innovation through dynamic capabilities, a model may prove to be useful.

Disruptive Innovation Model

Numerous studies have been conducted regarding the importance of innovation. For example, the Council on Competitiveness (2005) concluded that "Innovation will be the single most important factor in determining America's success through the 21st Century." In 2006, the American Management Association (AMA) commissioned a study on the emergence of innovation in global industries. The study concluded that "innovation is going to get considerably more important over the next decade"; therefore, it is essential for companies to eliminate the barriers of innovation and increase their innovative culture (AMA, 2006). IBM Global Business Services conducted an innovation study focused on public and private sector senior leadership. According to the study, CEOs expected fundamental changes for their organizations and saw opportunities to be seized through innovation (IBM, 2006). The study concluded that business model innovation and external collaboration are extremely important, as well as the role of senior leadership, in fostering an innovative climate. A study by the Boston Consulting Group found that the leading innovative organizations were characterized by risk-taking and investment in the long term (*BusinessWeek*, 2007; McGregor, 2007). The study also found that gimmick-driven campaigns were not the deciding factor – companies became innovative through hard work.

Innovative organizations are revolutionary in that they aggressively take markets from competitors (Hamel, 2002). Furthermore, innovation helps good organizations become great organizations and equips strong companies to become long-lasting entities (Collins, 2001). Additionally, resilient groups embrace disruptive change (Hamel and Valikangas, 2003), and competitive organizations use breakthrough ideas to destroy the opposition (Foster, 1986). However, the difficult challenge for most groups is creating an environment to foster breakthrough innovation while marginalizing practices that stifle creativity. While many business scholars have articulated innovation as a key for survival, deriving a formula for success has proven to be a difficult challenge. Throughout the literature though, there is evidence that motivation, focus, barriers, and culture play a crucial role in the emergence of breakthrough and game-changing ideas. By examining these key elements with regard to innovation, a base model for the emergence of disruptive innovation can be formulated. After presenting the model, implications for the defense industry will be briefly discussed.

Motivations for pursuing innovation

The primary reason companies pursue innovation is to gain and/or maintain competitive advantage. Foster (1986) explained that competitive advantage can only be achieved by going on the attack and that companies can lose their markets almost overnight to faster-developing technologies. Based on recent research and literature, several consistent themes appear among both industry professionals and corporate CEOs. As categorized below, the leading reasons for pursuing innovation within industrial organizations are to respond to customer demand, increase profitability, and improve efficiency.

Ranking of the quest for innovation (AMA, 2006)

1. To respond to customer demands
2. To increase operational efficiency
3. To increase revenues or profit margins
4. To develop new products and services
5. To increase market share
6. To better use new technologies.

Ranking of expanding the innovation horizon (IBM, 2006)

1. Profitable growth
2. Preempt business threats and create them
3. Drive needed efficiency
4. Develop multiple channels with different approaches for different customers.

Increasing profits

An increase in overall revenue and profit margins continues to be one of the primary motivations for companies to pursue innovation. The world's most innovative companies traditionally see greater revenue growth and margin growth compared to their less innovate counterparts (*BusinessWeek*, 2007). However, companies are finding it takes time to see profit growth and are often abandoning innovation investments for more short-term gains. Most decisions being made regarding innovation, and particularly the development of dynamic capabilities, would benefit from a long-term perspective.

Responding to customer demand

In today's marketplace, innovation is often seen as a primary means to acquire and hold onto customers. Peters (1997) explained this concept best: "If the other guy's getting better, then you'd better get better faster than the other guy's getting better, or you're getting worse." However, it is important to understand the level of customer interaction envisioned – while working closely with the customer provides great insight into their needs, it can also hinder the recognition of emerging needs and technologies (Francis and Bessant, 2005). Therefore, a high level of customer interaction seems to be more appropriate for sustaining/incremental innovation efforts, while disruptive/radical innovation typically requires less customer involvement.

Improving efficiency

As shown below, companies need to reduce cycle times and improve operational efficiency to survive. Hammer and Champy (2001) explain that because of customer power and customer choice, simply relying on acceptable process performance is no longer sufficient; furthermore, they state that conventional business remedies do not address the source of the problem, which is non-value-added work resulting from fragmented processes.

Cycle time reduction goals in industry (Defense Science Board, 2007)

Automobile: past (24 months); recent (24 months); goal (<18 months)
Commercial aircraft: past (8–10 years); recent (5 years); goal (2.5 years)
Commercial spacecraft: past (8 years); recent (18 months); goal (12 months)
Consumer electronics: past (2 years); recent (6 months); goal (less than 3 months).

In the space industry, recent R&D accomplishments by Space X corporation may upend the abovementioned benchmarks and set new standards for innovation trend lines.

Focus of innovation resources

While the need to focus resources on innovation is widely espoused, the optimal balance of investment is widely debated in the literature. Short-term investments necessitate close attention to detail, midterm investments demand capital and a willingness to take risks, and long-term investments require imagination and technological daring (Hayes and Abernathy, 1980). Innovation strategies by companies today are best described by looking at investments by functional area, innovation magnitude, and innovation type. The studies and literature indicate trends toward customer focus, reliance on business model innovation, and an emerging push toward new breakthrough products/services.

Customer focused innovation

According to the American Management Association (2006) survey results summarized below, more than 25% of the innovation resources in participating companies were focused on supporting customer experience and service. In addition, the study found that while innovation occurs across various functional areas, the areas directly related to customer relationships are receiving the highest degree of focus. Marketing, sales, customer service, and supply chain functions equated for over 41% of the functional areas of innovation.

Functional areas of innovation

R&D (27%)
Marketing (17.2%)
Information technology (12.2%)
Sales (9.7%)
Customer service (8.9%)
Manufacturing (6.5%)
Supply chain (5.4%)
Planning (5.1%)
Human resources (3.9%)
Finance (2.4%).

Focus areas of innovation

Customer experience (15.2%)
Service (12.6%)
Core processes (12.4%)
Product performance (12.2%)
Enabling processes (11.8%)
Business models (10.6%)
Brand (8.4%)
Networks and alliances (8.1%)
Product systems (4.7%)
Channel (3.6%).

Emphasis on business model innovation

Companies are finding with greater certainty that business processes and organizational innovation are important. The IBM (2006) study found that "four out of every ten companies were afraid that changes in a business competitor's business model would upset the competitive dynamics of the entire industry." It's no wonder then that the CEOs of outperformers are placing nearly twice as much focus on business model innovation than the CEOs of underperformers.

Product/service migration toward disruption

While competition has pushed companies to consider process innovation, the most popular type of innovation focus continues to be in the area of products/services. The recent industry shift is toward new products/services with "fewer companies focusing on incremental innovation or making minor changes to existing products" (*BusinessWeek*, 2007). This further solidifies the importance of understanding the emergence of disruptive innovation.

Barriers of innovation

Innovation can be a difficult and daunting challenge – one of the reasons for this is that most innovation experts agree that barriers hampering innovation are abundant. Many companies invest considerable resources into fostering ideas only to have their innovation efforts squelched by internal and external barriers (Kelley and Littman, 2001). Below is a summary of the most common barriers to innovation found in companies today. Although the semantics of obstacles varies from study to study, several common themes are consistent throughout the research: unsupportive culture, insufficient resources, lack of strategic vision, and poorly developed processes, even in the most innovative companies (*BusinessWeek*, 2007).

- Lengthy development times
- Lack of coordination
- Risk-averse culture
- Limited customer insight
- Poor idea selection
- Inadequate measurement tools
- Lack of ideas
- Marketing or communication failure
- Insufficient resources
- Lack of formal strategy for innovation
- Lack of clear goals and priorities
- Unsupportive organizational structures
- Short-term mindset.

Internal
- Unsupportive culture and climate
- Limited funding for investment
- Workforce issues
- Process immaturity
- Inflexible physical and IT infrastructure
- Insufficient access to information.

External
- Government and other legal restrictions
- Economic uncertainty
- Inadequate enabling technologies
- Workforce issues arising externally.

Unsupportive culture

The research findings reveal unsupportive organizational cultures to be significant obstacles to innovation growth. This is consistent with Kelley and Littman's (2001) observation that company mindset is one of the biggest barriers to innovation. Risk adversity, inflexibility, communication failures, workforce issues, and lack of ideas are all common symptoms of a poor innovative culture. Overcoming these barriers can best be addressed by cultivating a positive innovative culture. The characteristics of innovative culture are addressed in more detail later in the chapter.

Insufficient resources

Innovation is not merely about financial investments – it also involves investments in people, facilities, markets, training, and technology. Many organizations are falling into the "performance" trap where the company is doing well and fails to explore other opportunities because of the time, money, and personnel required (AMA, 2006). Other organizations are opting to sacrifice long-term stability for short-term gains. With reductions in discretionary dollars and pressures from stockholders, many CEOs are forced to divert R&D resources to low-risk investments with guaranteed returns (IBM, 2006). According to the *BusinessWeek* (2007) assessment, "More than half of all CEOs, chairmen, and presidents of companies were happy with how they'd spent on growth initiatives. CFOs, not surprisingly, were among the least satisfied: A full 63% were unhappy with their results." This mindset clearly defines the difficulties faced by innovators attempting to gain access to needed resources.

Lack of strategic vision

Although it is debated in the literature whether companies can "direct" innovation, it is commonly acknowledged that innovation strategy plays a role in fostering new concepts. Based on the American Management Association (2006) research, most companies fall dramatically short in developing a well-understood strategy for innovation and a shared vision on how to execute a plan for innovation. Industry's lack of innovation strategy is summarized below based on a survey of respondents (American Management Association, 2006):

> Have a shared definition of what innovation is: 41.3%
> Regularly review the progress of innovation: 22.4%
> Have a shared agenda to execute the innovation strategy: 12.3%
> Have a well-understood strategy for innovation: 12.1%
> Have well-defined roles and responsibilities: 11.3%.

Poorly developed processes

Long development times, insufficient access to information, poor idea selection, ineffective organizational structures, and communication failures are all indicative of poorly developed processes. Hammer (1996) contends that "it is not uncommon to find less than 10 percent of the activities in a process to be value-adding, with the rest of the rest mostly non-value-adding overhead." Process improvement is based on a commitment to optimize value through a process view of accomplishing work. It is not surprising that companies with inefficient processes struggle with innovation given that it takes creative and radical thinking to develop effective processes.

CHARACTERISTICS OF INNOVATIVE CULTURE

Organizational culture is defined as "a system of shared meaning held by members that distinguishes the organization from other organizations" (Robbins and Judge, 2007). An innovative culture is, therefore, a shared organizational environment designed to foster innovation. Many companies even specialize in teaching organizations to become more innovative. IDEO, ranked as the 28th most innovative company in the world (*BusinessWeek*, 2007), is considered a premiere leader in the development of the breakthrough spirit. With the recent emphasis being placed on innovation throughout the business world, it is not surprising that hundreds of articles and publications have been written on the characteristics of an innovative culture. Several common threads appear within the leading studies that help define the key characteristics: strong customer focus, collaboration, effective processes, creative people, inspiring leadership, risk-taking, and motivation/reward systems.

- The right organizational structures
- The right processes
- The right people
- Inspired leadership
- Orchestration from the top
- Collegial culture with individual rewards
- Consistent business and technology integration
- Customer focus
- Teamwork and collaboration with others
- Appropriate resources
- Organizational communication
- Ability to select the right ideas for research
- Ability to identify creative people.

Strong Customer Focus

The research suggests that organizations who place their existing and future customers at the forefront tend to be more innovative. Strong customer focus does not just mean delivering what customers ask for but rather "capturing their ideas or actually allowing them to innovate on their own behalf" (American Management Association, 2006). According to Kelley and Littman (2001), co-founder of IDEO, true understanding comes not by talking to customers but by watching them and becoming immersed in their environment. As a result of this strong customer focus, organizations are in a better position to implement disruptive product and process innovations that transform the marketplace and decimate the competition. Demonstrating this point, Christensen and Raynor (2003) reviewed the extensive market analysis conducted by a quick-service restaurant chain with regard to milkshake sales. The group examined not just what the customers wanted, but why they wanted it, when they wanted it, who they were with, and what they would be doing if they were not there buying a milkshake. They essentially focused on the job the customer was trying to get done.

Collaboration

External and internal collaboration is a common characteristic found in studies on innovation. According to Hargadon (2003), most significant innovations come from collaborative groups of people and not brilliant lone individuals. Collaborative innovation can be defined using the organizational Garbage Can Model (Cohen et al., 1972). The theory articulates that many solutions to problems can often be found by sifting through garbage in which ideas, or the ideas of others, have been tossed out as being irrelevant. Similarly, innovative cultures are best characterized by broad and often unrelated people that simply interact to make breakthroughs happen. Organizations that collaborate to a large extent typically perform better than the competition and receive strong benefits from the innovate spirit that is generated.

Efficient Processes

Efficient processes are streamlined and provide the appropriate level of performance to the organization. In addition, efficient processes undergo an endless cycle of improvement in which performance is measured, benchmarks are established, gaps are identified, and modifications are implemented (Hammer, 1996). According to the American Management Association (2006) assessment, innovative cultures are strongly tied to how efficiently organizations can capitalize on ideas. Innovative organizations know how to balance resource investments, select the right ideas, mobilize the right resources, and measure results. The level of disruptive innovative is directly related to an organization's ability to get funding and manpower required to cultivate new idea proposals (Christensen, 1997).

Creative People

Creative people, a key element in creating an innovative culture, solve problems by examining the world from different perspectives (Glover and Smethurst, 2003). Innovators are able to look beyond the status quo and visualize the realm of the possible while not allowing risk and adversity to hamper their progress. Henry Ford reportedly once said, "Failure is the only opportunity to begin again, this time more intelligently." Not everyone is naturally creative, and many companies like IDEO have developed a series of innovation roles that allow people to contribute to the innovative culture. Although business scholars believe that innovation comes from groups of creative people, breakthrough teams are composed of individual characters and diverse personalities deliberately recruited to generate energy and ideas (Kelley and Littman, 2001).

Inspiring Leadership

Collins (2001) found that successful leaders, those who blend extreme personal humility with intense professional will, were the catalyst in building great companies. Supportive leadership has been shown to be an equally important characteristic in building an innovative culture. The extent to which the leader reflects on organizational objectives, strategies, and processes, and implements changes accordingly is directly related to the organizational climate for innovation. In organizations with more reflective leaders, employees rated the innovative climate higher, organizational practices were more nontraditional, and there was a greater amount of change (Kazama et al., 2002).

Risk-Taking

"Innovation demands adherence to two fundamental principles: a willingness to accept risk and a willingness to wait for the return on investment" (Council on Competitiveness, 2005). While most scholars agree that innovation is a risky venture, only 20% of global companies actually recognize and reward intelligent risk-taking (American Management Association, 2006). Innovative cultures are made stronger by embracing failure as an option and taking the time to experiment. IDEO describes this innovation characteristic with the slogan, "Fail often to succeed sooner" (Kelley and Littman, 2001). Encouraging risk-taking helps create an environment where employees are willing to take chances with radical ideas.

Motivation and Reward Systems

Rewards for innovative behavior were a common characteristic cited in several publications on innovative culture in industry. Most companies use nonfinancial rewards as a means to promote innovation (American Management Association, 2006). Companies that "reward individual [innovation] contributions achieved 2 percent higher operating margins on average and grew nearly 3 percent faster than those who did not" (IBM, 2006). Motivation and reward systems are closely tied with organizational willingness to accept risk.

How you encourage and reward innovative activities will ultimately determine whether your employees undertake them. Innovation starts with employees willing to take risks. Employees will be apprehensive of these activities if they perceive the upside to be limited and the downside to be significant. A truly innovative culture needs to make employees feel secure enough to believe that failure itself will not affect their position within the firm.

Deloitte (2003)

PUTTING IT ALL TOGETHER

Christensen and Raynor (2003) propose that building an organization capable of disruptive growth requires a careful balance of resources, processes, and values. Combining these thoughts with previous studies of organizational innovation provides a model for fostering disruptive innovation. The model proposes the following: an increase in the right motivation, plus an increase in the right focus of innovation resources, plus a decrease in the barriers of innovation, plus an increase in the characteristics of innovative culture, will foster an increase in the emergence of disruptive innovation. This model, summarized below, is not intended to be an equation for guaranteed success but rather a conceptual formula to ensure that critical elements in the emergence of disruptive innovation are considered. While the interpretations, applications, and considerations will be domain dependent, the basic model is a universal framework for innovation improvement. Understanding the model is not sufficient though – to link it back to the previous discussion, fostering innovation also requires a full understanding of the organization's processes and its capabilities.

INNOVATION GUIDE FOR BUSINESS AND INDUSTRY

- View disruptive innovation as constructive development
- Embrace innovation revolution strategically
- Foster an atmosphere of innovative engagements

- Understand your motivations for innovation
- Focus resources on innovation pursuits
- Facilitate and support innovative culture
- Make concerted efforts to eliminate or mitigate innovation barriers
- Support breakthrough risk management.

DEFENSE IMPLICATIONS

The suggestions hitherto, along with the concepts of processes and dynamic capabilities, are applicable to all organizations in both the public and private sectors. To show the applicability to the defense community, each element of the recommendations is briefly discussed in this section of the chapter.

DEFENSE MOTIVATIONS FOR PURSUING INNOVATION

Within the defense community, the reasons for pursuing innovation may be quite different and caution must be exercised. For example, defense organizations could be motivated by a desire to be viewed as state of the art and capable of using new technologies more effectively. From a dynamic capabilities perspective, this could reflect a perceived need to enhance the organization's assets and improve its technological positioning. However, it could also be an indication of focusing too much on incorporating the newest technologies to create a "wow" factor. It could also indicate a reliance on technology, and perhaps a focus on invention instead of innovation, to meet the customers' needs. Depending on the situation, a better approach may be to focus on the job the customer is performing and strive to help the customer perform that job better (e.g., more quickly, more effectively, less costly).

Relying on policies to encourage innovation may not be very effective. Tidd (1993) found that policies often do not support technology strategies; instead, organizations tend to either follow industry trends or act in an ad hoc manner in response to a near-term need. This type of reactive approach may be due to existing learning mechanisms (or the absence thereof); therefore, defense organizations interested in innovation might consider examining their organizational processes to ensure appropriate structures and polices are in place to develop congruent strategies. Recognizing the impact of past strategic choices, coupled with critical thought about the impact of current decisions on future opportunities, could also be helpful. Since Chesbrough and Rosenbloom (2002) consider the business model to be a mediator between technology and value, the defense community might consider placing more emphasis on the business model aspect of innovation and developing appropriate value propositions. Finally, processes should be in place to facilitate cross-functional teamwork and integration, as well as to introduce employees to new technologies and make them aware of their potential uses and benefits; this learning process thus affects the evolutionary path of the organization.

The key factor is whether processes are in place to address the motivations for pursuing innovation. To facilitate this desire, it is important that organizations consider all components of the dynamics capabilities framework and develop commensurate strategies. With a narrow focus instead of a broader perspective, organizations may be overlooking opportunities to improve their dynamic capabilities and be more innovative.

The functional and focus areas of innovation were presented earlier in this chapter. The low percentages shown could indicate an overall weak application of innovation efforts; however, it could also reflect a lack of focus. When this happens, Francis and Bessant (2005) suggest that innovation efforts often develop without any coherent strategy and are often inefficient and sometimes contradictory. They also suggest that systematic analysis and comparative benchmarking might help facilitate more alignment between incongruent innovation efforts. A more structured approach to the development of organizational processes could also be helpful.

Networks and alliances are a key source of innovation (von Hippel, 1988) in which the primary reason for collaborating is to access either complementary technologies to support innovation activities or new markets (Tidd, 1993; Greis et al., 1995). However, too much focus on collaborations (i.e., networks and alliances) could reflect a reliance on external entities to drive innovation efforts instead of developing organic capabilities. Additionally, collaborations can affect an organization's evolutionary path by potentially shaping, sometimes positively and sometimes negatively, future strategic choices (Teece, 2006). Therefore, defense organizations are encouraged to develop a healthy strategy toward the use of collaborations.

When it comes to the customer experience, business models and brand areas are important components. Since business models help convey the organization's value proposition (Chesbrough and Rosenbloom, 2002), the choice of business model will influence the organization's processes, positions, and paths (Francis and Bessant, 2005). Therefore, more emphasis on innovative business models could potentially provide new benefits for defense organizations. Although branding may not be very applicable in the defense community, it may be helpful in establishing effective communication channels with customers to provide a better understanding of what innovation can do for them.

DEFENSE BARRIERS TO INNOVATION

Consistent with the RBV of the organization and other research (e.g., Blumentritt and Danis, 2006), defense organizations often indicate that insufficient resource is a primary barrier. However, Liao et al. (2009) found that the primary constraint hindering innovation is the lack of integrative capabilities (e.g., routines for integrating external knowledge and identifying opportunities). What this tells us is that organizations tend to lack processes to perform the coordination/integration, learning, and reconfiguration roles which Teece et al. (1997) claim are necessary to develop new competencies quickly.

Furthermore, a lack of guidance from the organization's leadership may suggest that innovation is accomplished in an ad hoc manner. Employees may feel they are getting adequate support from their immediate supervisors but not receiving clear guidance from the organization's senior leaders. The defense community may, thus, benefit from examining strategies and guidance since the ability of senior managers to "sense and seize" opportunities while overcoming organizational inertia and path dependencies is at the core of dynamic capabilities (O'Reilly and Tushman, 2007). This is especially important since organizational constraints are often "hidden" in everyday activities and processes. A frequently overlooked constraint is the organization's history and the path-dependent nature of capabilities created by the organization's routines. As previously mentioned, these core capabilities can easily become core rigidities (Leonard-Barton, 1992).

Finally, a "fear of failure" culture is a potential innovation barrier. Employees may not relate individual attitudes to barriers; however, when viewing culture as a barrier, they may be thinking of the organization's processes, policies, and procedures. This may be why factors related to organizational culture – threat of new ideas, lack of rewards, and short-term mindset – are often rated higher than the "fear of failure" barrier. In some ways then, culture may be viewed in terms of bureaucracy, which Francis and Bessant (2005) characterize as unfriendly to innovation.

CHARACTERISTICS OF DEFENSE INNOVATIVE CULTURE

Of primary concern to the defense community may be the freedom to innovate, which may be because of the bureaucratic and structured nature of most government organizations. This is consistent with the SAB's (2006) finding that the Air Force relies too much on technology demonstrations instead of experimentation. To be truly disruptive, Christensen and Raynor (2003) suggest the use of discovery-driven planning, to include experimentation and learning. An innovative culture also requires appropriate organizational processes and leadership ability to reconfigure assets and "sense and seize" opportunities. This may be lacking in government organizations, thus making the culture not as conducive as it could be in terms of facilitating innovation. Additionally, defense leaders may want to ensure there is a clear understanding, shared definition, and strategy for innovation in their organizations.

Industry considers the best way to establish an innovative culture is to focus on the customer. Although the defense community may consider customer focus to be important, it may struggle with the degree of "connectedness" to the customer and efforts to develop an appropriate value proposition and business model. Another important factor for industry is effective organizational communication, which requires effort and supporting processes. Therefore, poor communication may contribute to it being seen as a barrier to innovation. It may also imply more of a team approach to developing innovative solutions as compared to the typical "stovepipes" in more bureaucratic organizations.

DEFENSE LEADERSHIP INVOLVEMENT

Although defense organizations may consider innovation to be extremely important, they may find that it is not integrated very well into the overall organizational strategy. Blumentritt and Danis (2006) have suggested that "strategic orientation may be a powerful explanatory variable that accounts for important differences in how innovation is managed." In fact, de Jong and Marsili (2006) found that there is a correlation between the presence of a documented innovation strategy and the level of innovative activity in an organization. Similarly, O'Reilly and Tushman (2007) recommend that leaders articulate a vision and strategic intent, along with identifying specific complementary organizational processes. Furthermore, Lawson and Samson (2001) found that innovation often requires visionary leadership; coordination between innovation, business, and technology strategies; and a commitment to results. Therefore, the defense community may want to consider using strategy to facilitate the integration of innovation. They may also find it helpful to develop new value propositions and business models.

SUMMARY AND RECOMMENDATIONS

Managing innovation creates a dilemma for organizations. A loose organizational structure is often perceived as flexible and, thus, preferred if one wants to foster innovation, creativity, and adaptability. However, a formal structure and key management controls are required to coordinate and communicate innovation efforts. The key is to have a broadly structured framework within which employees have the freedom to make decisions about the best approach to take for a specific effort. At a minimum, each organization should have a tailored version of the innovation funnel. The intent of the funnel is to generate ideas, narrow the list of ideas to those that are most promising, and then implement the ideas that are selected to increase the value provided to the customer.

If an organization wants to become more innovative, the following principles are offered for consideration.

1. Create a strategic vision that establishes innovation as a priority.
2. Inspire the workforce by clearly identifying the organization's challenges and discussing how innovation will help address those challenges. Keep in mind that innovation is not required in every organization.
3. Evaluate the organization's dynamic capabilities and determine the changes required to align them with the strategic vision. Successful innovation depends on two key factors – resources and capabilities. Does the organization have the appropriate resources? Does the organization have the appropriate dynamic capabilities?
4. Review the organization's existing processes and create/change processes as required. This includes the innovation process itself, as well as complementary processes within the organization. Determine how innovation will be integrated with other processes in the organization.

5. There's an old adage in organizations – "you get what you measure." Therefore, spend some time developing an effective set of metrics to measure innovation and communicate the results.
6. Innovation is accomplished through people. Therefore, provide training to the workforce in terms of product and/or process innovation tools, managerial tools, and general problem-solving skills.
7. Recognize innovative behavior and reward innovative results.
8. Promote experimentation and prototyping as a way to develop a "fail early and often" mindset.

Although innovation is rooted in curiosity and discovery, it's not freewheeling and void of structure – it's driven by a system of principles and practices which support and encourage people to solve problems. Therefore, and as previously mentioned, innovation should be considered a process – a process that can be managed. It's ultimately a management and leadership question involving choices to be made about resource allocation and coordination. With the right choices and the proper approach to developing dynamic capabilities, the military can position itself to fulfill the following vision expressed by Gen "Hap" Arnold at the end of World War II.

The next war may be fought by airplanes with no men in them at all... Take everything you've learned about aviation in war, throw it out of the window, and let's go to work on tomorrow's aviation. It will be different from anything the world has ever seen.

REFERENCES

Air Force Studies Board. (2016). *The Role of Experimentation Campaigns in the Air Force Innovation Life Cycle*. Washington, DC: The National Academies Press.

American Management Association and Human Resource Institute. (2006). *The Quest for Innovation: A Global Study of Innovation Management 2006–2016*. New York: American Management Association.

Barney, J. (1991). Firm resources and sustained competitive advantage, *Journal of Management*, 17(1), 99–120.

Blumentritt, T. and Danis, W. (2006). Business strategy types and innovative practices, *Journal of Managerial Issues*, 18(2), 274–291.

BusinessWeek. (2007). Special Report - 2007 Most Innovative Companies. Retrieved 7/25/2007, from www.buisnessweek.com/innovate/di-special/20070503mostinnovative.htm.

Chambers, J. (Ed.) (1999). *The Oxford Companion to American Military History*. New York: Oxford University Press.

Chesbrough, H. and Rosenbloom, R. (2002). The role of the business model in capturing value from innovation: Evidence from Xerox Corporation's Technology Spinoff Companies, *Industrial and Corporate Change*, 11(3), 529–555.

Christensen, C.M. (1997). *The Innovator's Dilemma*. Boston, MA: Harvard Business School Press.

Christensen, C.M. and Raynor, M.E. (2003). *The Innovator's Solution*. Boston, MA: Harvard Business School Press.

Cohen, W. and Levinthal, D. (1990). Absorptive capacity: A new perspective on learning and innovation, *Administrative Science Quarterly*, 35, 128–152.

Cohen, M.D., March, J.G., and Olsen, J.P. (1972). A garbage can model of organizational choice, *Administrative Science Quarterly*, 17(1), 1.

Collins, J. (2001). *Good to Great*. New York: Harper Collins Publishers Inc.

Council on Competitiveness. (2005). Innovate America: Thriving in a World of Challenge and Change.

Deloitte Touche Tohmatsu. (2003). Fostering and innovative culture: Sustaining competitive advantage, *Growth: The Executive Series for Dynamic Companies*, 10(1), 7–24.

Defense Science Board. (2007). *2006 Summer Study on the 21st Century Strategic Technology Vectors: Volume IV Accelerating the Transition of Technologies into U.S. Capabilities*. Washington, DC: Defense Science Board.

de Jong, J. and Marsili, O. (2006). The fruit flies of innovations: A taxonomy of innovative small firms, *Research Policy*, 35, 213–229.

Drucker, P. (2002). The discipline of innovation, *Harvard Business Review*, 80(8), 95–102.

Eisenhardt, K. and Martin, J. (2000). Dynamic capabilities: What are they? *Strategic Management Journal*, 21(10/11), 1105–1121.

Foster, R. (1986). *Innovation: The Attacker's Advantage*. New York: Simon & Schuster Inc.

Francis, D. and Bessant, J. (2005). Targeting innovation and implications for capability development, *Technovation*, 25, 171–183.

Glover, C. and Smethurst, S. (2003). Creative license, *People Management*, 9(6), 1.

Greis, N., Dibner, M., and Bean, A. 1995. External partnering as a response to innovation barriers and global competition in biotechnology, *Research Policy*, 24, 609–630.

Hamel, G. (2002). *Leading the Revolution*. Boston, MA: Harvard Business School Press.

Hamel, G. and Valikangas, L. (2003). The quest for resilience, *Harvard Business Review*, 81(9), 52.

Hammer, M. (1996). *Beyond Reengineering: How the Process-Centered Organization Is Changing Our Work and Our Lives*. New York: Harper Business, A Division of Harper Collins Publishers.

Hammer, M. and Champy, J. (2001). *Reengineering the Corporation: A Manifesto for Business Revolution*. New York: Harper Business, A Division of Harper Collins Publishers.

Hargadon, A. (2003). *How Breakthroughs Happen: The Surprising Truth about How Companies Innovate*. Boston, MA: Harvard Business School Press.

Hayes, R.H. and Abernathy, W.J. (1980). Managing our way to economic decline, *Harvard Business Review*, 58(4), 67.

IBM Global Business Services. (2006). Expanding the Innovation Horizon: The Global CEO Study 2006.

Kazama, S., Foster, J., and Hebl, M. (2002). Impacting Culture for Innovation: Can CEOs Make a Difference? *In 17th Annual Conference if the Society for Industrial and Organizational Psychology*, Toronto, Canada.

Kelley, T. and Littman, J. (2001). *The Art of Innovation*. New York: Random House Inc.

Lawson, B. and Samson, D. (2001). Developing innovation capability in organizations: A dynamic capabilities approach, *International Journal of Innovation Management*, 5(3), 377–400.

Lee, H. and Kelley, D. (2008). Building dynamic capabilities for innovation: An exploratory study of key management practices, *R&D Management*, 38(2), 155–168.

Leonard-Barton, D. (1992). Core capabilities and core rigidities: A paradox in managing new product development, *Strategic Management Journal*, 13, 111–125.

Liao, J., Kickul, J., and Ma, H. (2009). Organizational dynamic capability and innovation: An empirical examination of internet firms, *Journal of Small Business Management*, 47(3), 263–286.

Matthews, J. and Shulman, A. (2005). Competitive advantage in public-sector organizations: Explaining the public good/sustainable competitive advantage paradox, *Journal of Business Research*, 58, 232–240.

McGregor, J. (2007). The World's Most Innovative Companies. *BusinessWeek*, Retrieved 7/25/2007, from www.businessweek.com/innovate/content/may2007/id20070504_051674.htm?chan=innovation_special+report+--+2007+most+innovative+companies_2007+most+innovative+companies.

Morison, E.E. (1966). *Men, Machines, and Modern Times*. Cambridge, MA: The MIT Press.

O'Reilly, C. and Tushman, M. (2007). Ambidexterity as a Dynamic Capability: Resolving the Innovator's Dilemma. Working Paper 07-088.

Pablo, A., Reay, T., Dewald, J., and Casebeer, A. (2007). Identifying, enabling, and managing dynamic capabilities in the public sector, *Journal of Management Studies*, 44(5), 687–708.

Peters, T. (1997). *The Circle of Innovation: You Can't Shrink Your Way to Greatness*. New York: Vintage Books, A Division of Random House Inc.

Porter, M.E. (1985). *The Competitive Advantage: Creating and Sustaining Superior Performance*. New York: Free Press.

Prahalad, C. and Hamel, G. (1990). The core competencies of the corporation, *Harvard Business Review*, 68(3), 79–91.

Robbins, S.P. and Judge, T.A. (2007). *Organizational Behavior*. Upper Saddle River, NJ: Pearson/Prentice Hall.

Rumelt, R. (1984). Towards a strategic theory of the firm. In: *Competitive Strategic Management*, Lamb, R.B., (Ed.) Englewood Cliffs, NJ: Prentice Hall, pp. 557–570.

Thal, A.E. and Shahady, D.E. (2019). Is your organization ready for innovation? In: *Defense Innovation Handbook*, Badiru, A.B., Barlow, C., (Ed.) Boca Raton, FL: CRC Press/Taylor & Francis Group, pp. 189–207.

Teece, D. (2006). Reflections on "profiting from innovation", *Research Policy*, 35, 1131–1146.

Teece, D., Pisano, G., and Shuen, A. (1997). Dynamic capabilities and strategic management, *Strategic Management Journal*, 18(7), 509–533.

Tidd, J. (1993). Technological innovation, organizational linkages, and strategic degrees of freedom, *Technology Analysis and Strategic Management*, 5(3), 273–284.

Tidd, J. and Bessant, J. (2009). *Managing Innovation: Integrating Technological, Market, and Organizational Change*. Chichester: John Wiley & Sons.

United States Scientific Advisory Board. (2006). Report on System Level Experimentation. No. SAB-TR-0-602.

von Hippel, E. (1988). *Sources of Innovation*. New York: Oxford University Press.

Innovation, Quality Engineering, and Statistics

7

INTRODUCTION

Adapted and reprinted by copyright permission from:
George E. P. Box & William H. Woodall (2012) Innovation, Quality Engineering, and Statistics, *Quality Engineering*, 24:1, 20–29, DOI: 10.1080/0 8982112.2012.627003 (a Taylor & Francis Group publication)
This chapter is an adaptation of the 2012 journal publication by Box and Woodall (2012), in which a linkage was made between innovation, quality engineering, and statistics. The contents of the original paper bear out the benefit of approaching innovation from a diverse systems viewpoint. The chapter has historical significance because, as long ago as 2012, the interest in the ramifications of innovation was already being pitched. This shows that the present wave of innovation is not totally new. Like all cycles of hype, innovation is here now. It may subside over time, but it will rise again in the future.

The chapter highlights the roles of statistics and quality engineering in the innovation process in business and industry. We review approaches that can be used in order to increase the chances of innovative discoveries. Most importantly, we stress the necessity for the quality engineering community to strengthen and promote its role in innovation. As Bisgaard (2006) has said, we should reframe much of what we do as systematic innovation. Adapting to the changing business and economic climate can revitalize our profession; failure to adapt threatens it.

There has been much discussion in the business literature on the importance of innovation and how businesses can best manage innovation. Thus, Michael E. Porter, Harvard Business School professor and leading management expert, has pointed out, "Innovation is the central issue in economic prosperity." As Porter and Stern (2001, p. 2) said, the way companies can gain a business advantage today is "to create and commercialize new products and processes, shifting the technology frontier as fast as their rivals can catch up."

Although primarily known for his emphasis on quality, W. E. Deming (1993, p. 10) said, "The moral is that it is necessary to innovate, to predict needs of the customer, give him more. He that innovates and is lucky will take the market."

Quality and efficiency cannot compete against the right innovation. Voehl (1995) recounted Deming's story of the buggy whip manufacturer who had a highly efficient process with excellent quality but whose company collapsed because of the failure to foresee and adapt to the horseless carriage. Christensen (1997) described a number of more recent cases in which successful companies fell from the top of their market because of their failure to adapt to the innovative technology that was competing against them.

In this chapter, we discuss how the use of statistics and quality engineering can aid in the innovation process. We provide a brief introduction to innovation, realizing that the topic is quite broad. We discuss some approaches to innovation and give some examples of how statistics is being used for innovation in industry. Most importantly, we implore the quality engineering community to reframe much of its work as innovation and give greater emphasis to this area.

WHAT IS INNOVATION?

Bisgaard (2006) defined innovation as the complete process of development and eventual commercialization of new products and services, new methods of production or provision, new methods of transportation or service delivery, new business models, new markets, or new forms of organization. Thus, innovations can occur in marketing, investment operations, and management techniques as well as in manufacturing and services.

Breakthrough innovation and incremental innovation are commonly used terms. Breakthrough innovations are often associated with new products or services and incremental innovations with improvements in current services or products. Unless an invention or idea is commercialized and produces a profit stream, it is not considered to be a useful innovation regardless of its novelty or its promise.

Innovation is, of course, not new but has sometimes been neglected. Though there are many famous historic examples, we pick just one. At the end of World War II, Japanese industry was in ruins. If you visit the Toyota museum in Japan, you will be shown their first car, an exact copy of a Volkswagen. There were many innovations that subsequently brought their cars to world attention. Three of these were (1) previously unknown standards of quality control, (2) new designs developed with the help of thousands of statistically designed experiments, and (3) an attitude toward the workforce based on the idea that they were all one family dedicated to making a good product, with the workers treated fairly by management. Toyota also introduced lean manufacturing techniques, as discussed by Womack et al. (1990). US manufacturers were slow to adapt to these innovations.

A primary purpose of quality engineering is to promote innovations. Some companies are now hiring "innovation engineers" with job descriptions quite similar to those for "quality engineers." There is interest in innovation within the quality engineering community. American Society for Quality (ASQ) has a technical committee on innovation. Several recent publications include Bhalla (2010), Montgomery (2008, 2011), Scriabina (2011), and Erto (2009).

It is widely acknowledged in the innovation literature that the vast majority of innovations are combinations of existing ideas and technology. Ingenuity may be required to form an effective combination, but none of the components themselves have to be new. For example, it is the effective combination of existing approaches and methods that made Six Sigma one of the most successful innovations in quality engineering, a point also made by Ron Snee and Roger Hoerl in Jensen (2011).

Sometimes an innovative system can be created using combinations of statistical tools that can add significant value to a business on an ongoing basis. These systems may involve combining several sources of data and extensive use of information technology, as in some of the "statistical engineering" applications discussed by Snee and Hoerl (2011). Golek (2011), for example, described a system for updating credit scoring models using credit card data, retail bank data, and external data such as that from credit bureaus. This system allowed more frequent updating of customer information and credit sources. Information was automatically routed to the appropriate business units, resulting in better financial decision-making. The system also allowed for more effective process monitoring and easier experimentation. Systems such as this one can improve analytical capability throughout an organization.

INNOVATIONS IN STATISTICAL SCIENCE

We can use an example in statistical science to illustrate the innovation process and how innovations can be combinations of existing components. In the area of process monitoring, Walter A. Shewhart (1931) proposed the key ideas of taking observations over time, distinguishing between common and special causes of variation, detecting process changes, and controlling the number of false alarms. Subsequent major advances were Western Electric's (1956) *Statistical Quality Control Handbook*, which operationalized his methods, the cumulative sum chart (Page, 1961), and the exponentially weighted moving average chart (Roberts, 1959). Innovations in statistical process control (SPC) have resulted primarily from combining these key ideas with other areas of statistics, including time series methods, process control, multivariate analysis, nonparametric methods, change-point methods, and operations research, among many others.

Many of the SPC methods are based on the assumption of independence and stationarity of the observations over time, conditions that are often inadequate approximations to reality. When a nonstationary approximation is necessary, Box and Narasimhan (2010) showed that variation about the target value can be minimized by allowing for the non-stationarity and using charts that also show when process adjustment is needed.

Some of the most recent research in SPC involves functional data analysis, image analysis, and spatiotemporal methods, reflecting the increasing amount of data now available. Much more data are likely to result from initiatives such as that by the MTConnect Institute, which has the goal of making operating data from machine tools easily accessible. There is a considerable opportunity in quality engineering for the more effective use of image data. As discussed by Megahed et al. (2011), images are becoming cheaper to obtain and can contain a considerable amount of information on process and product quality.

A primary driving force in the development of statistical science has been the need to adapt and invent theory and methods to handle practical problems faced by scientists and other practitioners. Developments in statistical science have also been needed to adapt to the increasing amounts of data available in many applications. Increases in computing power have made the analysis of large sets of data and the use of computationally intensive methods possible.

The innovative use of statistics in data-intensive applications was recently discussed by Pantula (2011) in his American Statistical Association (ASA) Presidential Address. There also has been an increasing use of data visualization, illustrated in a very entertaining and informative way by Rosling (2006).

Other growing fields are health informatics and the area of business analytics, described by Davenport and Harris (2007), in which companies use data mining to gain a competitive advantage.

APPROACHES TO INNOVATION

Three important approaches to innovation are inductive–deductive learning, lateral thinking, and group discussion. They are not rivals. One should not ask which is better. They are often best used in combination. The book by Box et al. (2005) gives many examples of innovation. The following is an extract from the introduction.

Inductive–Deductive Discovery

With the inductive–deductive discovery process, learning progresses iteratively, moving from idea to data, from data back to idea, and so on. An idea can be in the form of a model, hypothesis, theory, or conjecture. Data are the observations of measurements, facts, or phenomena.

The iterative inductive–deductive process between model and data is not esoteric but is part of our everyday experience. As one example, an engineer parks his car every morning in an allocated parking space. One afternoon he is led to the following inductive–deductive learning sequence:

Model: Today is like every day.
Deduction: My car will be in its parking space.
Data: It isn't.
Induction: Someone must have taken it.
Model: My car has been stolen.
Deduction: My car will not be in the parking lot.
Data: No, it is over there.
Induction: Someone took it and brought it back.
Model: A thief took it and brought it back. Deduction: My car will have
been broken into.
Data: It's unharmed and unlocked.
Induction: Someone who had a key took it.
Model: My wife used the car.
Deduction: She probably left a note.
Data: Yes. Here it is.

Box et al. (2005, p. 2) added the following:

> The iterative inductive–deductive process, which is geared to the structure of the human brain and has been known since the time of Aristotle, is part of one's everyday experience.
>
> Suppose you want to solve a particular problem and initial speculation produces some relevant idea. You will then seek data to further support or refute this theory. This could consist of some of the following: a search of your files and of the Web, a walk to the library, a brainstorming meeting with coworkers and executives, passive observation of a process, or active experimentation. In any case, the facts and data gathered sometimes confirm your conjecture, in which case you may have solved your problem. Often, however, it appears that your initial idea is only partially right or perhaps totally wrong. In the latter two cases, the difference between deduction and actuality causes you to keep digging. This can point to a modified or totally different idea and to the reanalysis of your present data or to the generation of new data.

Humans have a two-sided brain specifically designed to carry out such continuing inductive–deductive conversations. Though this iterative process can lead to a solution of the problem, one should not expect the nature of the solution to be obvious or the route by which it is reached to be unique. Often it is when things do not happen as expected that we learn. Thus, Isaac Asimov wrote, "The most exciting phrase to hear in science, the one that heralds new discoveries, is not 'Eureka!' (I found it!), but 'That's funny…'"

The inductive–deductive innovative discovery process is also evident in the Shewhart–Deming Plan–Do–Check–Act cycle. Its use in sequential experimentation was illustrated, for example, by Box and Liu (1999). They used the sequential design of a paper helicopter as a way to experience the innovation process. They begged others not to simply reanalyze their data or repeat their experimental runs. To learn to innovate, one must provide ideas and an apparatus of one's own.

Use of Lateral Thinking

The inductive–deductive process is, of course, not the only way to discovery and innovation. With de Bono's (1970) lateral thinking, one solves a problem not by working further down the established inductive–deductive path but by finding a new direction. The disadvantage of inductive–deductive thinking is that the scheme you have arrived at may have occurred to competing scientists and engineers who have similar educations and are working with the same set of scientific principles. This is less likely to happen with lateral thinking.

The lateral thinking concept is easier to demonstrate than to define. A simple example of lateral thinking concerns the quandary of a person who has to organize a tennis championship. Supposing that there are 47 contenders, how many matches would be necessary to come up with a winner in a single elimination tournament? The answer could be obtained by enumeration, but it can be reached much more easily by thinking about the losers. There have to be 46 losers, so this is the number of needed contests. A good statistical example of de Bono's lateral thinking occurred at a Princeton seminar where Mervin Muller discussed a way of generating normal deviates by piecewise approximation of the normal curve. This was complicated and messy and it seemed that there ought to be a simpler way. This led Box to the question "What is it in the normal distribution that is uniformly distributed?" For two independent normal deviates, the angle of the radius vector and the log of its length are distributed uniformly and independently. So, this provides a way of generating pairs of independent normally distributed random variables from pairs of random numbers. (The less than two-page note containing this result (Box and Muller, 1958) has been cited almost 1,400 times on Google Scholar (www.scholar.google.com).)

These applications of lateral thinking are not earth shattering, but the idea can be. This was demonstrated, for example, by Charles Darwin. Everyone could see how wonderfully each living thing fit into its environment, so it was argued that this must be the result of intelligent design. Darwin, thinking laterally, realized that all that was needed was reproduction and natural selection. (As an exception to the general rule, evolution was almost simultaneously discovered by A. R. Wallace.)

Another example of lateral thinking was R. A. Fisher's use of n-dimensional geometry in statistical analysis and the design of experiments. This led, at once, to his ideas of degrees of freedom, orthogonality, the additive property of independent sums of squares, important properties of univariate and multivariate normal distributions, and the idea of sufficiency. It also led to the analysis of variance, the development of regression analysis, and a better understanding of Gauss's method of least squares. W. A. Jensen (W. L. Gore & Associates, personal communication, 2011) commented the following:

> One of the things that I've found that helps spur lateral thinking is to take a step back and look at the big picture. The engineer may be down in the details of how to design a particular experiment for some predetermined purpose or how to solve a particular issue. If I come back to the business context and understand what the overall goals are of the project or team, we

may discover that the predetermined purpose may actually need to change, which changes the necessary details. I find myself often asking questions "What is the goal of your experiment? What are the questions you are trying to answer? Why do you want to answer those questions? What will you do if you get this answer?" Keeping the big picture in mind has proved to be extremely successful for me personally. This is not something I learned in school but have learned from many different consulting situations.

Lateral thinking is counterintuitive and will usually be resisted. It is easy to understand this. We have all been trained to think in a certain way when faced with a problem and one automatically calls on all that he has been taught and has experienced. Thus, at first, Darwin's ideas were indignantly rejected, as were Fisher's.

Use of Discussions and Group Interaction

A third approach uses cross-functional discussion teams. The discovery process can be greatly catalyzed by group discussion if the group contains people from different disciplines. Adair (1990) has discussed how teams should be formed and run in order to be most effective. Scholtes et al. (2003) described the many aspects that can make this method effective.

Discussion groups are important not only in and of themselves but also as a necessary adjunct to the other approaches. Thus, for example, de Bono's (1985) "six thinking hats" method can be regarded either as a means to facilitate lateral thinking or as a way to facilitate discussion within groups.

It is important for there to be openness and trust on teams; otherwise, potentially useful ideas may not be suggested. Many point out that with an experienced team, there is little distinction between work and play. Many of these characteristics of successful teams were evident in the many years of the "Monday night beer session" held in the basement of the home of George Box. There was no official connection of these sessions to the University of Wisconsin nor were there attendance or course requirements. Students and faculty from many departments, and sometimes people from industry, came and talked about their problems in an atmosphere in which ideas could be openly exchanged and discussed. Many found these sessions rewarding and productive. For example, University of Waterloo Professor Bovas Abraham wrote, "I can testify that I learned more about statistics in those sessions than what I got in standard courses" (Abraham, 2010).

The lack of distinction between work and play on teams leads to a very positive work environment. Berkun (2010, p. 105) quoted Lewis Thomas, former dean of the Yale Medical School, as saying, "One way to tell when something important is going on is by laughter."

THE USE OF STATISTICS AND EXPERIMENTATION

In this section, we use some examples to illustrate the use of statistics and quality engineering in the development of innovations. Market research studies are used to identify opportunities for innovation. Statistical thinking and methods are often required in the design and analysis of the measurement systems needed with new products and processes. The extensive use of experimentation and prototyping is an important aspect of innovation.

Many problems are complicated and contain many variables of interest. Experimentation is usually expensive. Statistical experimental design and, in particular, fractional factorial designs can minimize cost and maximize effectiveness. In addition to manufacturing, experimentation is now being widely applied in business.

Thus, Tang et al. (2010) wrote, "At Google, experimentation is practically a mantra; we evaluate almost every change that potentially affects what our users experience." Statisticians and engineers at Google use live Internet traffic experimentation to test new features, such as the background color of ads. They believe that their methods can be used by others interested in improving search engines and other Web applications.

Bell et al. (2006) described how marketers have embraced experimentation for marketing and advertising testing. Factorial designs and Plackett–Burman designs have increased the speed, power, and profitability of the testing programs. Capital One is very well known for the use of this type of experimentation to determine the most effective ads and promotions in their credit card business.

Bullington et al. (1993) reported the use of fractional factorial experiments in the manufacturing process for industrial thermostats at Eaton Corporation. As a result of their experimentation, the average life span of the thermostats, measured in on–off cycles, was increased by a factor of 10 with some of the process factor levels set at less expensive values. The performance of the longer-lasting and more robust thermostats that resulted allowed the penetration of new markets.

In 1969, chemical engineer Robert W. Gore ran a series of experiments to try to stretch heated rods of polytetrafluoroethylene (PTFE) by about 10%. After a series of failed experiments in which the heated rods were stretched slowly, he applied a sudden forceful yank that caused a rod to stretch about 800%. This counterintuitive result led to the raw material ePTFE, which became the foundation for many of the products of the now $3 billion company

W. L. Gore & Associates. This example illustrates that it can pay to vary factor levels beyond what may be considered reasonable based one's intuition about a process. For more information on Gore's experimentation, see Chemical Heritage Foundation (2010).

An area of increasing importance is computer experimentation, described by Santer et al. (2003). Due to the, sometimes, excessive costs of building physical prototypes, a computer model of the process is constructed. Experimentation then occurs with the computer model. The importance of computer experimentation in innovation was discussed by Thomke (1997, 2003), who reported, for example, how BMW was using virtual experiments for crash testing in order to lower costs and speed development. Testing with physical prototypes was done for confirmation after the virtual testing. Procter and Gamble uses computer experimentation in research and development where the idea is to "explore digitally, but validate physically" (Lafley and Charan, 2008, p. 200).

Another approach used to accelerate the pace of experimentation in the discovery process is high-throughput screening (HTS). This is an automated method for experimentation used for drug and compound discovery in the chemical and pharmaceutical industries. HTS allows researchers to quickly conduct thousands of chemical or pharmacological tests. The goal is to identify active compounds that warrant further study. Malo et al. (2006) discussed statistical issues related to HTS.

INNOVATION AND SIX SIGMA

Some argue that a focus on quality stifles innovation. For example, Johnson (2011, p. 148) wrote, "Innovative environments thrive on useful mistakes and suffer when the demands of quality control overwhelm them." Similarly, Hargadon (2003, p. 22) wrote, "Breakthrough innovations don't usually mix well with the pursuit of six sigma quality control nor with those customers who just purchased your last generation of products." Similar views are given in the oft-quoted *Business Week* article by Hindo (2007). For example, Dartmouth Professor Vijay Govindarajan was quoted as saying, "The more you hardwire a company on total quality management, (the more) it is going to hurt breakthrough innovation."

We do not understand why an emphasis on quality and the use of Six Sigma is a barrier to innovation. If one introduces a new product or service, there will be quality problems to solve, but quality tools and Six Sigma can be helpful in solving them. One can work to reduce the number of errors and failures in manufacturing and services while still allowing for failures in innovation efforts. One should not try to manage all aspects of a business in the same way. Managing quality involves reducing risk, whereas in managing innovation, one must allow significant risk and be open to unavoidable failed efforts. Innovation management includes promoting creativity, the generation and evaluation of ideas, using a gateway system to narrow down the ideas to the most promising ones, the extensive use of experimentation and prototyping for testing, and eventually production and marketing for the best ideas. There are numerous issues in innovation management that are not included in quality management. Thus, the American Society for Quality's (ASQ, 2010) statement that "It makes sense to manage innovation activities with the same management tools and approaches that are used in other major sectors of the business" seems mistaken.

Design for Six Sigma (DFSS) methods can be used to develop both incremental and breakthrough innovations. Inductive–deductive iteration in the form of the Define–Measure–Analyze–Improve–Control (DMAIC) process improvement framework can also lead to innovations. Hoerl and Gardner (2010) discussed the role that lean Six Sigma plays in innovation and creativity.

In the early use of Six Sigma, a major purpose was for defect reduction. Later, process efficiency was emphasized. Now Six Sigma methods are also used for developing new products and services that reach new and broader markets; that is, for innovation. Montgomery and Woodall (2008) referred to these three eras of Six Sigma use as Generations I, II, and III, respectively.

It is unfortunate that so many still associate Six Sigma only with defect reduction. Perhaps this is a by-product of the Six Sigma 3.4 ppm metric and the associated distributional assumptions, including the allowance for a 1.5 sigma variation in the process mean and the use of specification limits. As Montgomery and Woodall (2008) argued, this metric is a nonessential aspect of the Six Sigma process improvement and product design frameworks and is now doing more harm than good.

FURTHER COMMENTS

Leadership

The importance of leadership cannot be overemphasized. It is true that many people helped Thomas Edison develop the light bulb, many sailors helped Admiral Lord Nelson win the Battle of Trafalgar, and no doubt many engineers and scientists helped Steve Jobs develop the iPhone. Nevertheless, these happenings would not have occurred (at least not at that time) without these leaders. One reliable guide to effective leadership is that of Scholtes (1998).

General Advice to Individuals

Much has been written on how individuals can become more creative and thus become more able to identify and solve problems. Many experts recommend that individuals keep their ideas in notebooks or on a computer. In a busy work environment, ideas can easily be forgotten if not written down. It helps to return to these past ideas after a period of time when one may have a new perspective.

All those offering advice to individuals point out the necessity of hard work and persistence. Persistence certainly plays a big role. There is nothing less obvious than the obvious. It is best if you try to forget everything you know and try to think of a problem from first principles (Peña, 2001).

An encouragement for innovation is to put oneself in a position in which, in order to solve a particular scientific problem, one is forced to learn and discover new things. Thus, response surface methodology was developed to meet a daily challenge to improve chemical engineering processes at Imperial Chemical Industries (ICI; Box, 1954; Box and Wilson, 1951). Later at the University of Wisconsin–Madison, there was a 3-year effort in cooperation with the Chemical Engineering Department to design and build a reactor that was self-optimizing (Box and Chanmugam, 1962; Kotnour et al., 1966). This produced unexpected benefits. It resulted in a better understanding of time series and of dynamic systems and control in science and business. As discussed in Pen~a (2001), this work led to the book on time series analysis by Box and Jenkins (1970).

Use of Analogy

Another approach that can prove useful in innovation is the use of analogy. The use of analogy is a technique in Theory of Inventive Problem Solving (TRIZ), an idea generation approach now used by many companies (see Altshuller, 1996). As an example, at ICI, one way to improve processes was by running designed experiments. It was expensive and disruptive to run full-scale experiments, and small-scale experiments in the pilot plant could be misleading. The graphical representation of the imaginary evolution of a species of lobster was used to illustrate the idea of evolutionary operation to company executives. This statistical procedure made it possible to generate information in a manner analogous to evolution and natural selection to improve a product during actual manufacture. Under evolutionary operation, relatively small changes based on simple experimental designs are made to a process close to normal operating conditions. One is thus able to move process factors toward better settings. This procedure also has the capability of following moving maxima. For more information, see Box (1957). Evolutionary operation was named as one of the earliest examples of evolutionary computation, now a widely used computer technique (Fogel, 1998).

RECOMMENDATIONS FOR THE QUALITY PROFESSION

Innovation is looked upon far more favorably by business leaders than quality. Although quality will remain important, it is clear that breakthrough innovation can trump quality improvement. Bisgaard (2006) and Bisgaard and De Mast (2006) made the case that quality engineering needs to be expanded and much of it repositioned as innovation engineering. Bisgaard's (2006) article, which is reprinted in this issue of Quality Engineering, said that in order to take advantage of the opportunities and benefits of this reframing of quality engineering, "We need to rethink our professional activities, our conferences, our educational materials, our textbooks, our role in the educational system both in engineering and business schools, and how we manage our journals." Many of these changes would be straightforward, requiring us simply to recognize the work that is already being done. A step in this direction was the addition of the word innovation into the title of the second edition of the text on designed experimentation by Box et al. (2005).

Bisgaard (2006) reiterated that Six Sigma provides a systematic approach to process and product improvement and innovation and that design of experiments and computer experimentation are important tools within the product development process. Standard quality tools are also useful for incremental innovations that can be vitally important. We should shift the perception of Six Sigma away from defect reduction and deemphasize or abandon the Six Sigma 3.4 ppm metric. We need to emphasize Generation III Six Sigma, not Generations I and II. We believe that ASQ needs to change its approach toward innovation. The field of quality control and improvement is closely tied to innovation and new product development, but the quality area is not viewed by business leaders as being related to innovation. ASQ (2010) argued that quality and innovation have much in common in correcting the notion that quality and innovation are at odds with each other. Given these perceptions, it seems wise that ASQ move aggressively to embrace the field of innovation. If meaningful changes are made within the organization, there would be advantages for ASQ to be renamed ASQI. ASQ will always be associated with quality, not innovation, without a change in its name.

We recommend that our conferences have a greater focus on innovation. As an example, the annual Quality and Productivity Research Conference (QPRC; see www.qprc2012.com/) rarely includes any presentations on what would be considered "production research" found, for example, in the *International Journal of Production Research*. A more accurate name for the conference would be the "Quality and Innovation Research Conference (QIRC)." Topics related to innovation are currently featured at this conference but should be emphasized to a greater extent.

The two most highly rated industrial engineering journals (by *Journal Citation Reports'* Impact Factor) are the innovation-related journals *Technovation* and the *Journal of Product Innovation Management*. The editors of quality-related journals should be encouraged to put greater emphasis on topics related to innovation.

Given its importance and the high level of interest by the business community, we need to be bold in expanding our field to include what is becoming known as innovation engineering.

ACKNOWLEDGMENTS

The work of Professor Woodall was partially supported by NSF Grant CMMI-0927323. The authors greatly appreciate the helpful comments of Fadel Megahed, PhD candidate in the Virginia Tech Grado Department of Industrial and Systems Engineering, Sundar Dorai-Raj of Google, William R. Myers of Procter & Gamble, Frederick W. Faltin of the Faltin Group, and Willis A. Jensen of W. L. Gore & Associates.

REFERENCES

Abraham, B. (2010). George Box: A source of inspiration for quality and productivity, *Quality Engineering*, 22, 103–109.

Adair, J. (1990). *Leadership for Innovation: How to Organize Team Creativity and Harvest Ideas*. London: Kogan Page Limited.

Altshuller, G. (1996). *And Suddenly the Inventor Appeared: TRIZ, The Theory of Inventive Problem Solving.* (Translated by Lev Shulyak). Worcester, MA: Technical Innovation Center, Inc.

American Society for Quality. (2010). Fresh Thinking on Innovation and Quality. ASQ White Paper. Available at: www.asq.org (accessed October 20, 2011).

Bell, G.H., Ledolter, J., and Swersey, A.J. (2006). Experimental design on the frontlines of marketing: Testing new ideas to increase direct mail sales, *International Journal of Research in Marketing*, 23, 309–319.

Berkun, S. (2010). *The Myths of Innovation.* Sebastopol, CA: O'Reilly Media, Inc.

Bhalla, A. (2010). What's the big idea? Fostering innovation turns employees into problem solvers, *Quality Progress*, 43(6), 39–43.

Bisgaard, S. (2006). The future of quality technology: From a manufacturing to a knowledge economy & from defects to innovations. 2005 Youden Address, *ASQ Statistics Division Newsletter*, 24(2), 4–8. Available at: www.asq.org/statistics/ (Reprinted in Quality Engineering 24(1):29–35, 2012.)

Bisgaard, S. and De Mast, J. (2006). After Six Sigma—What's next? *Quality Progress*, 39(1), 30–36.

Box, G.E.P. (1954). The exploration and exploitation of response surfaces: Some general considerations and examples, *Biometrics*, 10, 16–60.

Box, G.E.P. (1957). Evolutionary operation: A method for increasing industrial productivity, *Journal of the Royal Statistical Society -Series C*, 6(2), 81–101.

Box, G.E.P. and Chanmugam, J. (1962). Adaptive optimization of continuous processes, *Industrial & Engineering Chemistry Fundamentals*, 1(1), 2–16.

Box, G.E.P. and Jenkins, G.M. (1970). *Time Series Analysis: Forecasting and Control.* San Francisco, CA: Holden-Day.

Box, G.E.P. and Liu, P.Y.T. (1999). Statistics as a catalyst to learning by scientific method, part I—An example, *Journal of Quality Technology*, 31(1):1–15.

Box, G.E.P. and Muller, M.E. (1958). A note on the generation of random normal deviates, *Annals of Mathematical Statistics*, 29(2), 610–611.

Box, G.E.P. and Narasimhan, S. (2010). Rethinking statistics for quality control, *Quality Engineering*, 22, 60–72.

Box, G.E.P. and Wilson, K.P. (1951). On the experimental attainment of optimum conditions, *Journal of the Royal Statistical Society-Series B*, 13(1), 1–45.

Box, G.E.P. and Woodall, W.H. (2012). Innovation, quality engineering, and statistics, *Quality Engineering*, 24(1), 20–29. DOI: 10.1080/08982112.2012.627003

Box, G.E.P., Hunter, J.S., and Hunter, W.G. (2005). *Statistics for Experimenters—Design, Innovation, and Discovery*, 2nd ed. Hoboken, NJ: John Wiley & Sons.

Bullington, R.G., Lovin, S., Miller, D.M., and Woodall, W.H. (1993). Improvement of an industrial thermostat using designed experiments, *Journal of Quality Technology*, 25(4):262–270.

Chemical Heritage Foundation. (2010). Robert W. Gore. Available at: www.chemheritage.org/othmerlibrary.org/discover/chemistry- in-history/themes/petrochemistry-and-synthetic-polymers/syntheticpolymers/gore.aspx (accessed September 2, 2011).

Christensen, C.M. (1997). *The Innovator's Dilemma.* New York: HarperCollins.

Davenport, T.H. and Harris, J.G. (2007). *Competing on Analytics: The New Science of Winning.* Boston, MA: Harvard Business School Publishing.

de Bono, E. (1970). *Lateral Thinking: Creativity Step by Step.* New York: Harper & Row.

de Bono, E. (1985). *Six Thinking Hats*. Boston, MA: Little, Brown and Company.

Deming, W.E. (1993). *The New Economics for Industry, Government, Education*. Cambridge, MA: Massachusetts Institute of Technology, Center for Advanced Engineering Study.

Erto, P. (Ed.) (2009). *Statistics for Innovation*. Milan: Springer-Verlag Italia.

Fogel, D.B. (Ed.) (1998). *Evolutionary Computation – The Fossil Record, Institute of Electrical and Electronics Engineers*. New York: Wiley-IEEE Press.

Golek, J.L. (2011). Statistical Engineering within Financial Services. In: *Paper presented at the Joint Statistical Meetings*, Miami, FL.

Hargadon, A. (2003). *How Breakthroughs Happen—The Surprising Truth about How Companies Innovate*. Boston, MA: Harvard Business School Publishing.

Hindo, B. (2007). 3M's Innovation Crisis: How Six Sigma Almost Smothered Its Idea Culture. *Business Week*, June:8–14.

Hoerl, R.W. and Gardner, M.M. (2010). Lean Six Sigma, creativity, and innovation, *International Journal of Lean Six Sigma*, 1(1), 30–38.

Jensen, W. (Ed.) (2011). Statistics to Facilitate Innovation: A Panel Discussion. ASQ Statistics Division. Special Publication. (Reprinted in *Quality Engineering* 24(1): 2–19, 2012.)

Johnson, S. (2011). *Where Good Ideas Come From: The Natural History of Innovation*. New York: Riverhead Trade.

Kotnour, K.D., Box, G.E.P., and Altpeter, R.J. (1966). A discrete predictor-controller allied to sinusoidal perturbation adaptive optimization, *ISA Transactions*, 5, 255–262.

Lafley, A.G. and Charan, R. (2008). *The Game-Changer – How You Can Drive Revenue and Profit Growth with Innovation*. New York: Crown Publishing Group.

Malo, N., Hanley, J.A., Cerquozzi, S., Pelletier, J., and Nadon, R. (2006). Statistical practice in high-throughput screening data analysis, *Nature Biotechnology*, 24(2):167–175.

Megahed, F.M., Woodall, W.H., and Camelio, J.A. (2011). A review and perspective on control charting with image data, *Journal of Quality Technology*, 43(2):83–98.

Montgomery, D.C. (2008). Does Six Sigma stifle innovation? *Quality and Reliability Engineering International*, 24, 249.

Montgomery, D.C. (2011). Innovation and quality technology, *Quality and Reliability Engineering International*, 27, 733–734.

Montgomery, D.C. and Woodall, W.H. (2008). An overview of Six Sigma, *International Statistical Review*, 76(3), 329–346.

Page, E.S. (1961). Cumulative sum control charts, *Technometrics*, 3(1), 1–9.

Pantula, S. (2011). The 2010 ASA Presidential Address - Statistics: A key to innovation in a data-centric world, *Journal of the American Statistical Association*, 106, 1–5.

Peña, D. (2001). George Box: An interview with the International Journal of Forecasting, *International Journal of Forecasting*, 17, 1–9.

Porter, M.E. and Stern, S. (2001). National innovative capacity. In: The Global Competitiveness Report 2001–2002, Sachs, J., Porter, M., Schwab, K., directors. New York: Oxford University Press, pp. 2–18. Available at: www.isc.hbs.edu/Innov-9211.pdf (accessed October 20, 2011).

Roberts, S.W. (1959). Control chart tests based on geometric moving averages. *Technometrics*, 42(1), 97–102.

Rosling, H. (2006). Hans Rosling Shows the Best Stats You've Ever Seen. Available at: www.ted.com/talks (accessed October 20, 2011).

Santer, T.J., Williams, B.J., and Notz, W.I. (2003). *The Design and Analysis of Computer Experiments*. New York: Springer-Verlag.

Scholtes, P.R. (1998). *The Leader's Handbook: Making Things Happen, Getting Things Done*. New York: McGraw-Hill.

Scholtes, P.R., Joiner, B.L., and Streibel, B.J. (2003). *The Team Handbook*, 3rd ed. Madison, WI: Oriel Incorporated.

Scriabina, N. (2011). Organize how you innovate, *Quality Progress*, 44, 16–22.

Shewhart, W.A. (1931). *Economic Control of Quality of Manufactured Product*. New York: D. VanNostrand.

Snee, R.D. and Hoerl, R.W. (2011). Proper blending – The right mix between statistical engineering, applied statistics, *Quality Progress*, 44, 46–49.

Tang, D., Agarwal, A., O'Brien, D., and Meyer, M. (2010). Overlapping Experimental Infrastructure: More, Better, Faster Experimentation. In: *Proceedings of the 16th International Conference on Knowledge Discovery and Data Mining*, pp. 17–26, New York: Association for Computing Machinery.

Thomke, S. (1997). *Enlightened Experimentation: The New Imperative for Innovation*. *Harvard Business Review on Innovation*. Boston, MA: Harvard Business School Publishing, pp. 179–222.

Thomke, S. (2003). *Experimentation Matters: Unlocking the Potential of New Technologies for Innovation*. Boston, MA: Harvard Business School Publishing.

Voehl, F. (1995). *Deming: The Way We Knew Him*. Boca Raton, FL: CRC Press/ Taylor & Francis Group.

Western Electric. (1956). *Statistical Quality Control Handbook*. Indianapolis, IN: Western Electric Corporation.

Womack, J.P., Jones, J.T., and Roos, D. (1990). *The Machine That Changed the World: The Story of Lean Production*. New York: Rawson Associates.

Index